Science and Fiction

For further volumes:
http://www.springer.com/series/11657

Science and Fiction – A Springer Series

This collection of entertaining and thought-provoking books will appeal equally to science buffs, scientists and science-fiction fans. It was born out of the recognition that scientific discovery and the creation of plausible fictional scenarios are often two sides of the same coin. Each relies on an understanding of the way the world works, coupled with the imaginative ability to invent new or alternative explanations—and even other worlds. Authored by practicing scientists as well as writers of hard science fiction, these books explore and exploit the borderlands between accepted science and its fictional counterpart. Uncovering mutual influences, promoting fruitful interaction, narrating and analyzing fictional scenarios, together they serve as a reaction vessel for inspired new ideas in science, technology, and beyond.

Whether fiction, fact, or forever undecidable: the Springer Series "Science and Fiction" intends to go where no one has gone before!

Its largely non-technical books take several different approaches. Journey with their authors as they

- Indulge in science speculation—describing intriguing, plausible yet unproven ideas;
- Exploit science fiction for educational purposes and as a means of promoting critical thinking;
- Explore the interplay of science and science fiction – throughout the history of the genre and looking ahead;
- Delve into related topics including, but not limited to: science as a creative process, the limits of science, interplay of literature and knowledge;
- Tell fictional short stories built around well-defined scientific ideas, with a supplement summarizing the science underlying the plot.

Readers can look forward to a broad range of topics, as intriguing as they are important. Here just a few by way of illustration:

- Time travel, superluminal travel, wormholes, teleportation
- Extraterrestrial intelligence and alien civilizations
- Artificial intelligence, planetary brains, the universe as a computer, simulated worlds
- Non-anthropocentric viewpoints
- Synthetic biology, genetic engineering, developing nanotechnologies
- Eco/infrastructure/meteorite-impact disaster scenarios
- Future scenarios, transhumanism, posthumanism, intelligence explosion
- Virtual worlds, cyberspace dramas
- Consciousness and mind manipulation

Nick Kanas

The Protos Mandate

A Scientific Novel

 Springer

Nick Kanas, M.D.
Professor Emeritus (Psychiatry)
University of California, San Francisco
San Francisco, California
USA

ISSN 2197-1188   ISSN 2197-1196 (electronic)
ISBN 978-3-319-07901-1   ISBN 978-3-319-07902-8 (eBook)
DOI 10.1007/978-3-319-07902-8
Springer Cham Heidelberg New York Dordrecht London

Library of Congress Control Number: 2014942134

Cover illustration: Uniformed couple on the observation deck of a starship gazing at a passing comet. © Mike Heywood

Springer is part of Springer Science+Business Media (www.springer.com)

Preface

The Protos Mandate is first and foremost a science fiction novel. It is a tale of the first group of people sent to an extrasolar planet to begin the process of expanding our species into the cosmos. It takes us from the building of the starship, through the long outbound period to reach the planet's star, Epsilon Eridani, to the landing and colonization of the planet itself, named Protos. The vicissitudes of this journey for the people involved, and what they discover on the planetary surface, forms the core of the story.

As suggested by the subtitle of this book, *A Scientific Novel*, the story will be followed by an appendix that reviews the current science and technology behind the story. This addition is one of the unique features of the science fiction stories that are part of the new "Science and Fiction" series being introduced by Springer Publications. I am proud to say that *The Protos Mandate* is my second novel in this series, the first being entitled *The New Martians*.

Trying to extrapolate the science and technology of the present into the future is difficult, since no one knows for sure what the future will be like several centuries from now. For example, where are the flying cars that were predicted several decades ago to be cluttering up our air-borne freeways? Nevertheless, we have robots and wrist telephones, so all was not lost! The best the present can do is to stimulate our interest and show us ways that the future could develop. It is up to us to actualize it.

The appendix will include a review of important science fiction stories over the years that have dealt with interstellar travel, of which there have been many. So, why write another such story? For many reasons. Much has been discussed in recent years about interstellar propulsion systems, psychological and sociological issues related to space travel, and the discovery of exoplanets, and it is time to incorporate the latest information about such issues into a novel that deals with a mission to a distant star. In addition, the story line explores an interstellar expedition from soup to nuts: from the origins of the mission, through its century-long flight phase, to the landing and colonization of a new world. It is very human-oriented, tracking the problems and vicissitudes of the people initiating the mission as well as their descendants. It also explores issues related to both suspended animation and multigenerational space travel, and what it might be like to have people cohabitating a starship with both experiences. This certainly increases the complexity of the mission! Finally, what might we expect to find on a distant planet that is similar to the way Earth was eons ago? No green men, but perhaps something else that is green (and brownish-yellow).

In writing The Protos Mandate, I want to thank a number of individuals whose help and influence contributed to its final publication. First are the staff members at Springer Publications, especially Dr. Harry Blom and Maury Solomon, who published the textbook I co-wrote with Dr. Dietrich Manzey entitled *Space Psychology and Psychiatry*. Special thanks go to Mr. Clive Horwood, the respected former publisher of Praxis Publications. Clive produced my two celestial cartography books under the Springer/Praxis label: *Star Maps: History, Artistry and Cartography* (now in its second edition), and *Solar System Maps: From Antiquity to the Space Age*. Clive put me in contact with Dr. Christian Caron, the co-editor of Springer's Science and Fiction series. He and his staff selected both *The New Martians* and *The Protos Mandate* to be part of this exciting new series.

I am grateful to Chris for his helpful comments to an earlier draft of this novel, along with the comments made by an anonymous member of the editorial board. I am also grateful for the useful comments made by a number of friends and colleagues: Drs. Ruth Corwin, Shirley Huang, Lyn Motai, Richard Ray, and my wife Carolynn, who has continued to support me in this and many other writing activities over the years. Of course, I am solely responsible for the ideas and concepts that appear in this book.

June 15, 2014   Nick Kanas M.D.
                                              Professor Emeritus (Psychiatry)
                                         University of California, San Francisco

Contents

Part I

The Novel

The Protos Mandate

I. LAUNCH

Image of a Ram-augmented Interstellar Rocket (RAIR) type of multigenerational starship moving through space. Through its forward-facing ramscoop, it collects interstellar hydrogen that is used as reaction mass to produce thrust. The hydrogen is energized by the fusion reaction of helium-3 and deuterium, which in this image is stored in tanks located behind the rotating wheel that houses and provides gravity for the crewmembers. Figure credit: © Steve Bowers (Orion's Arm Universe Project).

1. Prologue I

The rocket gleamed silver against the starry sky. Secured to its launch pad with trusses and clamps, it looked like a permanent extension of the desolate construction site. The site was built on a small rocky asteroid that had been moved decades earlier to its orbit around the Neptunian moon Triton. At the appointed time, the trusses silently separated and the clamps detached, accompanied by the ignition of the rocket's chemical engine. It strained upward to clear the site, then moved with gathering speed toward the faint laser beam ahead. The beam was generated from the Sun's light, collimated by the giant Fresnel lens orbiting Uranus, and directed toward the star Epsilon Eridani.

As the rocket entered the beam, it aligned itself in parallel and continued accelerating until its fuel ran out, at which point the chemical stage separated away into space. This left Probe EE-1 to continue the journey. At the front end of the two-meter long probe, a solar sail only a few millimeters in thickness began to unfurl, extending outward and outward until its three-kilometer diameter was reached. Pushed by the light beam, the probe began to accelerate toward its goal of 85 % light speed. When this cruising speed was attained, the Fresnel lens would be redirected to point the light beam toward another destination for another probe. Explosive charges would detach the solar sail into deep space so that its massive size would not interfere with the sensors that would be activated when Probe EE-1 arrived at its destination.

Some fifteen years later, upon reaching the Epsilon Eridani system, the probe's deceleration rockets would ignite to slow it down to a speed that would allow examination of the star's planetary system. The data would arrive back to Earth ten and a half years later, and the probe would continue on into the infinity of space. With any luck, one of the rocky planets known to orbit in Epsilon Eridani's habitable zone would have liquid water and an Earth-like atmosphere, temperature, and gravity, unlike the planets already examined around stars closer to Earth's Solar System. If deemed suitable, this remote planet would be targeted as the site of the first interstellar colony.

There was hope that the favorable conditions existing on such a planet would allow for the presence of life forms more complicated than the primitive forms found on Mercury, Mars, and the moons Europa and Titan. Whether or not intelligent life would be found on such a planet revolving around another star was actively debated by astrobiologists, but the consensus was that if the conditions allowed for a human colony to survive without expensive terraforming, then some sort of life would certainly be possible, if not likely. Time would tell.

2. Agwar

As the lunar shuttle lifted from the San Francisco District Space Port, Agwar Cinat looked out the window at the hazy scene below. As far as he could see through the polluted air, the skyscrapers extended south, interspersed here and there with the towers of the fusion reactors. The dike system holding back the rising waters of the Pacific Ocean and the inner Bay provided a clear outline of the region. This was not the case farther inland, where the inhospitable deserts extended toward the Rockies.

Agwar reflected on the richness and relative comfort of the region as compared with the conditions in his native state of Central Africa. Despite worldwide technological advances in climatology and agriculture that mitigated some of the climate changes produced by global warming, poverty continued to be widespread back home, the heat forced people to live indoors, and food production could barely keep up with the population of nearly three billion people. In contrast, the numerous fusion energy reactors, the desalination plants, and the genetically-engineered macro-farms and algae-supported micro-farms allowed the wealthy strip of land that extended from the San Francisco District to the San Diego District to support its 283 million inhabitants in relative comfort within their hermetically sealed and oxygen enhanced living structures. Interconnected up and down the coast like a massive spider web, these structures soared hundreds of stories into the sky and almost as deep into the depleted Earth. Indeed, the San-San Region was much more livable than the bordering Phoen-Houst and Mexico Regions, although none were as comfortable as the state of Canada, whose northern location kept the ambient temperatures under 50 degrees Celsius and moderated the titanic wind storms and tornadoes that were a fact of life elsewhere. In addition, the absence of polar ice had created a string of ice-free ports along the northern coastlines of Canada and Russia, which together with the mineral wealth helped support the 843 million inhabitants of the two states in relative comfort.

Agwar's reverie was interrupted by the space plane's intercom: "All passengers—please prepare for rocket drive."

His seat automatically reclined and the back cushion inflated as the shuttle accelerated upward and he was forced back into the cushion. As the angle of ascent increased, the seat pivoted to accommodate the *g*-force. After a few minutes, the force diminished as the murky sky gave way to the blackness of space and a carpet of stars. The Moon lay ahead.

I hope the Senate meeting is more civil this time, he thought. *Wu Cheng and Johann Schmidt will likely get into another verbal sparring match and make my life more difficult!*

Indeed, the Chinese and Martian Senators were always competing for power at the monthly 3-S Senate meetings, and lately he seemed to find himself in the middle of their outbursts.

I guess that's part of the job, he reflected further.

Project Protos was his main responsibility these days, and this put him in competition with several of the Solar System States. After all, trillions of dollars going into the *Protos 1* starship were taken out of the funds that could be used to relocate the billions of people crowded into the polluted regions of China or to assist the terraforming activities in the state of Mars. Furthermore, the relatively wealthy states of North America and Europe were having their own problems building dikes and desalination plants to support the billions of people in their own regions.

The Solar System is getting too crowded, the Earth too polluted … we just can't build enough space stations or hollow out enough asteroids for colonies or terraform Mars or the gas giant moons fast enough to take care of everyone, Agwar thought. *The stars must be our future!*

He found himself getting tense, and realizing he needed to conserve his energy, he took a Calm-doz tablet and drifted off to sleep. But his sleep was fitful and punctuated by dreams of people fighting with each other in blistering deserts.

****

Agwar awoke with a start to the announcement that the lunar landing was imminent. Glancing out the window, he saw miles of sealed buildings and their connectors, surrounded in the distance by a rim of craters.

There are even too many people here, he thought as he surveyed the city of 15 million.

The shuttle descended to the space port and landed with a slight bump, coming to a stop in front of the pressurized walkway that slid out to meet their airlock.

"Welcome to Luna City District Spaceport!" blared the voice on the intercom. "It is 1955 Solar System Universal time, and the date is September 30, 2444. We hope you enjoy your visit and fly again on Virgin Selene Spaceways."

Agwar stood up and walked out of the shuttle through the enclosed walkway that led into the spaceport. He spoke into his surgically implanted wrist compuphone and asked for a roboporter to retrieve his bag and summon a taxi. He got on the people-mover that went to the exit. He passed a number of stores and restaurants that were intermixed with corridors leading out to arrival gates bringing other passengers to the lunar capital from Earth, on-orbit stations, and various colonies in the Solar System. He also noticed a number of very tall, wiry people with lean muscles and long arms and legs.

Lunates, he thought to himself. *Nothing like growing up in 1/6th Earth gravity.*

With his Earth muscles, he felt buoyant but resisted the temptation to bound into the air in giant leaps. How would it look for the General Secretary of the Space Alliance to be playing air games on the Moon?

At the exit portal, a beautiful woman in a tailored jumpsuit awaited him with his bag. She had long auburn hair tied up in a bun, blue eyes, red lips, a gorgeous figure, and perfectly smooth skin.

A giveaway, he thought. *Skin without so much as a blemish screams 'robot'. The robotics industry here truly is the best in the Solar System.*

"Here is the bag that links to your wrist compuphone, sir. Can I assume it is yours?"

"Yes, thank you."

"No problem, sir. Your robotaxi is just outside. I hope you have a nice stay in Luna City."

Thinking how human-like her voice sounded, he entered the battery-powered vehicle and gave it directions to the Colony Hotel. It moved along almost silently within the pressurized connecting roadway to the nearest intersection, where it veered right into a large enclosed freeway. It then picked up speed with the other vehicles until turning again to first one, then a second, roadway that took him to the hotel. He got out, swiped his money card, and gave his bag to a waiting curbside roboporter who was very officious and dressed in a uniform with the red and gold colors of the hotel.

"Good day, sir. Welcome to the Colony Hotel. Do you have a reservation?"

"Yes I do, one room for three nights under the name of Cinat."

Pausing a moment to link with the reservation computer, the roboporter said: "Yes, sir, you are now checked in. All charges will be made to your money card. Let us know if you decide to stay longer. I will show you to your room."

"OK, thank you."

They walked by the stores and the bar/restaurant of the hotel into an outside-facing plexiglass elevator. As they ascended along the side of the hotel, Agwar marveled at the view of the city, with its gleaming metallic buildings and enclosed roadways. The Earth, visible in the brilliantly star-studded black lunar sky, appeared as a dusty blue and white marble through its polluted atmosphere. He glanced at his image that was reflected in the thick, pressurized glass: medium height and a bit plump, jet black skin, genetically-enhanced blue eyes, and thick white hair that mirrored his 72 years. Until the rigors of his job led to his divorce eight years ago, his ex-wife liked to run her hands through his hair, as did several of his recent partners, both male and female. He wondered if he would keep his hair when he got old, or if he would lose it as he entered the next century of his life.

The elevator stopped at his floor and they got out. As they approached his room, the roboporter turned on the voice activator unit and asked Agwar to speak clearly into the microphone to command the door to open. He did, and the now imprinted circular door silently enlarged like the iris of an eye, revealing a nice but standard room with a large bed, food center, voice activated holovideo monitor, bathroom, and glass-enclosed balcony.

Wonderful, he thought. *I have an Earth view.*

Placing the bag on the bed, the roboporter wished him a pleasant stay and departed as the door irised shut. Agwar unpacked, ordered a sandwich and scotch and soda from the foodbot, quickly imbibed both, and got ready for bed. Taking another Calm-doz tablet, he could not resist the impulse to leap into bed in the low gravity. He ordered the clock to wake him 9 hours later, at 0700 SSU time. Tomorrow he would be facing some confrontive senators, and he wanted to get as much sleep as he could.

3. Wu Cheng

Wu Cheng's arrival in Luna City was not very relaxed. He was met coming out of the shuttle walkway by embassy staff, hurried through the spaceport, and whisked directly by private car to the Chinese State Embassy, where the Ambassador awaited him. After some tea and cookies were promptly delivered, the two of them got down to business, switching their speech from Universal English to Mandarin.

"What has been the political opinion here the past few weeks about *Protos 1*?" Wu Cheng asked.

The Ambassador thought for a moment, then said: "The Luna State Senator and most of the other senators who reside here are in favor of the starship mission. The holovideo press and the polls of the populace of Luna City are also in support. Everyone seems to believe that colonizing exoplanets around other stars is worth the expense and that we have gone as far as we can in our Solar System. Ever since the Epsilon Eridani probe began sending back reports 29 years ago about the potential viability of Protos, people have seen the stars as a salvation."

The diminutive, prim 62-year-old Senator looked down at his manicured fingers and his immaculately tailored suit, then up at the Ambassador.

"I met with the Emperor before coming here, and he is most unhappy with the mission. He and the General Staff are very concerned with the unrest in our regions. The five billion people in the Beijing and Shanghai Regions alone are smothering in the pollution and heat. We barely have enough food to feed everyone, despite the increased imports from the Earth-orbiting hydroponic farms. The Emperor believes that we should conserve our resources, and he wants me to ask for additional relief funds at the 3-S Senate meeting tomorrow. This will be difficult. I expect other senators to point out that their states have also sacrificed tax money for the Protos expedition and that we need to solve our own internal problems ourselves. This seems to confirm what you have concluded here as well."

"Yes," said the Ambassador, putting down his tea cup as he scrunched his brow. "There is not much sympathy here for our cause. Relocating six billion people from the Regions in the east to the less populated western part of our state is an expensive endeavor. In addition, piping water from the desalination plants near the Sea of Japan to Mongolia and from the disappearing Himalayan snows to the Tibet Region are not politically popular activities in the eyes of our bordering states. You can expect the senators from Japan and India to be especially hostile tomorrow."

"And our recent war with the Taiwan State where the computers declared us to be the victor hasn't helped very much," Wu Cheng said. "In fact, the Emperor thought I should consider offering up Taiwan's sovereignty as a peace card to the Senate in return for cutting our taxes for Protos."

"That might work. As another small state, the people of Luna are very interested in the Taiwan cause and feel sympathy for its billion people. Some of the senators from other small states likely feel the same way. It might be worth a try."

"The problem is that the Reunification Party in China keeps pressing for immediate regionalization of Taiwan, obliterating its statehood, and securing control over its enormously productive hydroponic and algae farms. They would not relish our giving up our newly won possession."

After a pause, Wu Cheng stood up.

"Well, I will have much to ponder today. Thank you for the refreshments. Let me know if you have any further thoughts on the matter before tomorrow's meeting."

The two men shook hands and the Senator departed for his hotel.

As he walked back in the pressurized walkway, he thought about his family in Beijing. His second grandson was quite ill. He suffered from asthma, and the polluted city air only made it worse. At age 7, he regularly required oxygen from a portable dispenser, and no treatment had succeeded in quelling his disease. Likewise, Wu Cheng's brother, who was still a young man at 57, continued to suffer from congestive heart failure. This was made worse on high smog days when the temperature climbed to 60 degrees Celsius. He was pretty much restrained to living indoors, but the frequent energy blackouts disrupted the air conditioner and purifier machine, making the inside almost as intolerable as the outside. He wondered how much better life would be for all of them in the less populated west, away from the pollution and crowds and where the temperatures sometimes went down to a more tolerable 45 degrees.

The Emperor is right, he thought. *We must do something for our people right now, not in one or two hundred years when we colonize the stars. Diminishing our reliance on coal and fossil fuels has helped stem the tide, and fusion energy has proven to be relatively inexpensive and clean. But we need to do more. We don't need Taiwan or any other additional territory. We must stand up to the Reunification Party and others who would have us grow at the expense of our people's welfare. Fourteen billion people in China are enough. We need to pay attention to the now rather than to the uncertain future.*

Reaching his hotel, he went inside, went up into his room, and pulled up his remarks for tomorrow's meeting on his private computer. He read the holographic image of his speech, made a few corrections, then sent copies to his staff members for their final comments. He then changed his clothes and left his room for his next meeting.

4. Johann

Since his arrival four days earlier, Senator Johann Schmidt had attended a series of meetings with the mayor of Luna City, the Martian Ambassador, a group of businessmen and engineers working on the Terraform Mars project, and members of his staff. As he was walking away from his office the evening before the Solar System States Senate meeting, he marveled at the city around him.

It's nice having everyone enclosed in an interconnecting set of surface and underground buildings and passageways, he thought, *where the temperature and air are perfectly regulated throughout the city.*

In fact, only the people paying for Surface Safaris ventured outside to the lunar surface. For them, bounding around the airless and low-gravity Moon in their spacesuits was an exciting, if not somewhat dangerous, thrill, especially when they contracted to chase lunar lions hither and yon and shoot them with their laser rifles. Mounting a brain chip on their mantelpiece back home from a mechanized beast shot on the surface of the Moon was well worth the $ 300,000 price tag of the Safari.

Johann pondered his situation. When the Mars City Regional Council voted to take the next step in making the Martian atmosphere more breathable without the need for pressure masks, he took the proposal to the June 3-S Senate meeting last year. It was voted down, with the excuse that the masks were a minor inconvenience and that the trillions of dollars required to further improve Mars' atmosphere were needed to finish the construction of *Protos 1*. Especially galling was the Chinese Senator, who kept pushing for intra-state relocation money rather than funds to send his overflow people to Mars.

The fools could not see that by making Mars more like the Earth used to be, it would stimulate emigration from that horribly polluted place, he thought to himself. *With more people and resources, we Martians could expand towards the polar caps, closer to our water source, and we could move even closer to our goal of having a completely terraformed planet."*

He wasn't looking forward to the meeting tomorrow. He could make another appeal for the terraforming project, and likely Wu Cheng would ramble on about the Chinese relocation needs, but the votes were not there to transfer funds to these activities, since they would take the final launch money away from *Protos 1*.

He continued along the walkway, aware that people were staring at him. He was a striking figure: 52 years old, muscular, blond-haired, violet-colored eyes, tanned skin. Like other Martians, his appearance was surgically enhanced according to the latest styles. This vanity made his people both admired and scorned by others throughout the Solar System.

He looked through the transparent walls that enclosed him at the gleaming towers above and the openings to side streets and businesses along the way. Then, he became pensive.

No, another solution is indicated to the Protos situation, he thought, clenching his jaws as he considered his course of action.

5. Angelique

Angelique Moran worked feverishly in her office getting ready for the next day's meeting. Although she represented the State of Western Europe, she was also the Chair of the Senate Deep Space Committee, so she had been in Luna City for the past two weeks preparing for a possible showdown over the Protos project. She anticipated some resistance now that *Protos 1* was nearing completion, and she felt she might need to defend Agwar Cinat from the barking hounds of the 3-S Senate. She got up to take a bathroom break. Looking in the mirror, she saw a petite, attractive woman with a few uncorrected wrinkles around her 58-year-old brown eyes.

I must take care of those next week, she thought, dictating a reminder in her wrist compuphone to contact her plastic surgeon.

Anticipating a holocall from her husband, who was still in the Paris District tending to his robotics company, she applied some lipstick and fluffed her auburn hair. As if by magic, the holophone rang. She activated the visual, and a three dimensional image of her daughter Audrey appeared. The info window stated that she was calling from Space Station 3. Looking at her daughter was like looking at herself thirty years earlier.

"Hello sweetheart, how nice of you to call."

"Hello, mom. How are things going for your big meeting tomorrow?"

"Pretty well. I expect some controversy over the money being spent for the final push to launch *Protos 1*, but nothing I can't deal with. How is your work going?"

"It's pretty exciting. We're discovering more and more Earth-size rocky exoplanets in the habitable zones of stars. Several of them show oxygen in our computer-enhanced light spectra, so they'll be great candidates for future solar sail probes. I've been modeling biological life forms based on the spectra, and I'm coming up with some remarkable possibilities."

"Wonderful, dear. Your dad and I are very proud of you!"

Angelique had much to be proud of in her only child. The 32-year-old was one of the best astrobiologists in the Solar System, despite her relatively young age. Being based on Space Station 3 with its array of telescopes and equipment scanning the skies above the Earth's pollution was a real feather in her cap.

Fidgeting somewhat, Audrey stared directly into her mother's eyes.

"Mom, I have something to tell you. I've applied for one of the final Sleeper positions on *Protos 1*, and my application has been accepted. I'm going to the stars!"

Stunned, Angelique sputtered: "What! Really? I didn't know … I guess congratulations are in order, but do you really want to go? Suspended animation for over a century …"

"It is a bit scary. The procedure is relatively new, and there are risks that worry me. But the Sleeper route is the only way I can go to Protos, with the regular crewmember slots being all filled up. Plus, I'll be alive when the starship arrives at the planet. Like all of the other Sleepers, I have a skill that is not needed during the flight but will be on Protos. Think of it—as an astrobiologist, I'll have the chance to see and study alien life in another star system. It's an historic opportunity."

"Yes, but, we will never see you again, and when you arrive, your father and I will be long gone … "

"I know, mom, part of me really feels bad about this. I'll miss you too. And I know how much you and dad have always wanted grandchildren, and I would like this for you. But I really need to go. And you yourself have been so instrumental in this mission. I will be fulfilling your dream as well."

Angelique took pause at this. Her daughter was right that part of her wished she could go on the expedition herself after working so hard to support it after all of these years. And she and Jean-Paul had always wanted what was best for Audrey. But she loved her daughter and would miss her terribly, and there would be no grandchildren! She knew that her daughter had a mind of her own, and it would be fruitless to try and convince her to change it. Angelique hurt desperately inside, but she decided the best course would be one of support.

"You're right, dear. I guess I am being selfish. Of course it's a wonderful opportunity for you and your work. When do you leave for training?"

"In two weeks. I'll fly to the Protos Training Center in Valhalla City on Callisto for medical testing and interstellar travel orientation. This includes suspended animation psychological and physical testing to make sure that I'm not too nervous about the procedure or allergic to the various S-A chemicals and fluids that will be used. Then I'll have a month specialty training, which in my case means flying to Europa to study the organisms living in the subsurface water. It also means a trip to Jupiter to look for possible life floating in some of the layers of the atmosphere."

"Oh my, won't that be dangerous? I mean with the tremendous gravity pull of Jupiter, and the tremendous winds …"

"Yes, there'll be a risk, but the pilots assigned to our training are the best, and all have had success sliding into and out of the Jovian upper atmosphere."

"When do you meet the rest of the crew?"

"Right after the first of the year. I will be boarding *Protos 1*, where all of the crew will be quarantined until launch. Although we will have a chance to socialize a bit with everyone in the crew, we 40 Sleepers will be assigned to a separate part of the starship to receive further S-A orientation and really get to know one another for a few weeks. We'll be a special group. Apart from a few crewmembers who are leaving as young children, we'll be the only first generation colonists still alive when we reach Protos, the only ones who remember growing up on a Solar System body, the only ones who signed the Protos Mandate as adults and accepted the original purpose of the mission."

Angelique pondered what her daughter was saying.

"I suppose it will be interesting to wake up and find yourself with people who only know the experience of flying in space … people who only know about our Solar System and the reasons prompting the mission from historical records."

"Much of our training will deal with the psychology of this interaction between two quite different groups," Audrey said. "It'll be interesting for all of us."

"So, when will we see you again?"

"We'll be free the last two weeks of December, so I'll plan to see you and dad in Western Europe during the holidays."

"That will be lovely, dear. I think you should call your father and tell him your plans yourself. He too will be saddened but proud of your selection for the Protos mission."

"Thanks, mom. I knew you would understand."

The two of them continued talking about people, events, politics ... a number of things that were eclipsed by the Protos news. Then, with a promise to call her father in Paris right away, Audrey hung up.

Angelique went back to her work, but she couldn't concentrate. She feared that her husband would call her before their daughter reached him, and she would have to break the news herself. She was happy for her daughter but saddened that after the first of the year she would never see her again. She remembered her as a little girl, always curious and loving but restless, needing to go places and do things. Protos represented the fruition of these desires. Then she thought of the possibility that Audrey would never wake up from her S-A treatment.

Although freezing organs and other simple body parts had been done successfully for hundreds of years, putting entire people in suspended animation and reviving them safely after a long period of time was a different matter altogether. Different organs and body fluids had different freeze and thaw characteristics, and protecting the human brain from cognitive impairments and permanent memory loss was a particularly tricky procedure. After some disastrous failures with animals in the late twenty-first Century, research in this area was curtailed. In part, this was due to the technical difficulties. But like many other areas of science and technology, it took a back seat to high priority areas related to coping with global warming and overpopulation. Dikes had to be constructed in coastal areas to hold back the rising oceans; giant solar panels needed to be built in space to beam down the Sun's energy to collectors on Earth; fusion reactor technology had to be perfected to provide cheap energy that was safe and reliable in reactors located in urban areas; and hydroponic and desalination facilities needed to be built and widely distributed to provide food and water to the billions of people crowding the livable areas of the home planet. But when it became apparent that star travel would be necessary to deal with the teeming masses and resource depletion on Earth, research in suspended animation was revived. Despite a crash program, it was only in the past six years that a tolerable level of success had been achieved. However, the procedure had never been tried on anyone for more than three years. And being able to keep the cryogenic and life support equipment functioning at a safe and sustaining level for the entire 107-year mission had worried some of the Protos mission planners. This concern had been widely communicated by the media and was the primary reason that comparatively few people had volunteered to be Sleepers rather than regular crewmembers for this first interstellar expedition.

These thoughts made Angelique anxious. Leaving her work area, she got up and went into her bathroom, where she grabbed her bottle of Calm-doz.

Two tablets shortly relaxed her nerves, but they couldn't do anything about the tears that were streaming down her face.

6. The Leader

The hooded man walked quickly toward his evening meeting in a seedy part of Luna City. He passed a number of robogirls and roboboys who were provocatively dressed so as to advertise their wares. He marveled at how physically perfect they all were, even the women who were engineered with four large lactating breasts to satisfy everyone's mammary needs, and the men with huge bulges in their groins promising ample opportunities for enjoyment in their choice of three erection sizes. But he had no time for such activities today as he dismissively moved through the gauntlet of pleasure machines.

He soon found himself before the doorway to the Lunatic Lounge. He went in and asked the bodyguard for the private room. He was directed to the back, and on entering the room he saw his fellow conspirators. He was the last to arrive, but as leader of the group, the others had not yet conducted any business. They all ordered and imbibed a round of hashish oil with whiskey on the rocks, then began their meeting.

"So, I trust you all had a good trip here and were careful not to be followed?" began the Leader.

They all nodded in the affirmative.

"Good. We can't take any chances. Is everything in place?"

"Yes sir!" replied a dark, heavy-set bearded man from the regional colony on Enceladus. "My neighboring friends in the Titan colony are ready. Nobody will suspect them."

"That's the idea. With *Protos 1* being constructed around Callisto, the police will be looking for suspects in the state of Jupitermoon rather than Saturnmoon."

The petite bi-gender from Phobos leaned forward, adjusting his/her skirt and cracking his/her knuckles. "Are we sure that the Senate won't vote to delay the launch tomorrow? Now that the final decision to launch is at hand, the media has been reporting growing animosity among a number of special interest groups that want to delay the launch and divert launch money to Solar System solutions to Earth's problems."

"Yes," said the large balding man across the table, "We could build more orbiting satellites to off-load the people from Earth. My business associates have proposed a plan to construct an artificial ring satellite around the Earth that could hold millions of people. It would provide jobs here in the Solar System to match the terraforming activities taking place on Mars."

"Money could be made for us as well!" sneered the businesswoman from the Asteroid Titanium Company. "All of us in the Belt State would like to see more mining for materials to support these projects, but our Senator is unfortunately a small growth activist, not wanting to expand our industry so much that the rural character and small population of the outer Solar System will be destroyed."

"The Senate will not alter its course at this late date, with so much money already having been spent on *Protos 1*," intruded the Leader, scanning the room with penetrating eyes. "The people have been conned by the argument that colonizing Earthlike exoplanets will be more cost effective in the long run than building new orbiting cities or terraforming environmentally challenged planets and moons in our Solar System. They believe that in time less expensive and faster starships will be produced, and that a limitless number of Earth-like exoplanets will be found to accommodate humanity. We must stick to our plan to sabotage this fool's errand of a mission."

There was a general murmur of agreement. They all ordered another round of Hash n' Whiskey as they reviewed the plan. When they finished, each got up to go, leaving in groups of one or two so as not to arouse the suspicion of other people in the bar. Rather than reenter the sexual gauntlet, the Leader called for a robotaxi on his wrist compuphone and directed it to drive him away.

7. Senate Meeting

Agwar awoke the next morning feeling refreshed. After a light breakfast, he caught a robotaxi that whisked him to the center of Luna City. It stopped before a large gray building. Over the entrance, a holosign announced: "Solar System States Senate Building". He exited the taxi, and after passing though a retina security scan, he came to a bifurcation: the left corridor led to the administrative section, and the right to the Senate Assembly Hall, where he joined a procession of people who were all heading in the same direction. He greeted several of the senators who supported his program and avoided those in opposition. As he entered the hall, he was directed to the guest section in the front, where he took a seat. He faced a large semicircular array of desks occupied by the senators from all the states in the Solar System. There was a general buzz in the room as senators greeted each other or conferred with their staffs. Glancing at the agenda, he realized that the Protos mission was the first issue to be discussed.

"Order in the chamber," announced the booming voice of the Senate Secretary behind him. "Order!"

Everyone took their seats and activated their holographic computers.

"The October 2444 Senatorial Meeting of the Solar System States is now in session."

After a few general remarks and announcements, the Secretary began with the agenda.

"First on the docket is an update on the *Protos 1* starship. With us today is Mr. Agwar Cinat, the General Secretary of the Space Alliance. He will give us a progress report."

Agwar stood up and made his way to the podium at the focal point of the semicircle.

"Thank you, Mr. Secretary. Esteemed Senators and staff members. It is an honor for me to be here with all of you. I am pleased to report that the construction of *Protos 1* is on schedule and nearing completion in its orbit around Callisto. Using raw materials that have been collected and processed from the asteroid belt mines, the giant ramscoop has now been completed and is ready to collect hydrogen from interstellar space using a new design to minimize drag. This hydrogen will provide the thrust for the starship after it is energized by the fusion of helium-3 and deuterium. Nearly all of this fuel has already been transferred from the robotic mines in Saturn's atmosphere to four of the storage tanks around the *Protos 1* core. The other two tanks are full of start-up and emergency hydrogen for use when the starship travels through low hydrogen areas of space."

He paused to consult his notes, then continued.

"All of the crewmembers and the colony Sleepers have been selected, with crew training scheduled to begin in two weeks. We are still on schedule for a February 1, 2445 launch. Any questions?"

The light went on in front of Angelique's desk. She was recognized by the Senate Secretary.

"Mr. Cinat," she said as she stood up, "you and the entire Space Alliance are to be congratulated for keeping to the schedule. After all, the construction project has been going on for some 26 years, with only a few roadblocks and diversions that could have negatively impacted on the Protos mission. But you and your team have dealt with the problems and have persisted in building this truly marvelous vehicle."

"And we couldn't have done it without the help of the Senate Deep Space Committee, which under your chairmanship has been very supportive of our mission," he responded.

Agwar knew that he had scored some points when she smiled at being so recognized, which was especially important during this election year. But he meant what he said. Senator Moran was a space enthusiast, and for the duration of her tenure as the committee chair, she had managed to sequester the trillions of dollars that were needed for the project.

"I am pleased that you have finished picking the crew," Angelique said. "No doubt they will be well-trained for the mission, and careful attention will be given to their safety, especially the Sleeper group."

Aware that her daughter had been selected as a Sleeper, but not sure if Angelique had yet been told, Anwar played it safe: "Yes, Senator, safety is uppermost in our minds. I'm sure that many of you know that developing a successful suspended animation procedure has bedeviled our scientists for many years. After it was decided to go to Protos, a crash S-A program was instituted that finally yielded successful results, first in animals, then in humans. We are now confident that a reliable procedure has been tested to allow for a 106-year S-A duration, with no negative sequelae to the Sleepers after awakening."

"What are the final numbers for the crew composition?"

"We will start with 240 awake crewmembers and 40 Sleepers, Senator. According to the Protos Mandate, this number optimally uses available resources on the starship and should be maintained until it arrives at Epsilon Eridani and people shuttle down to Protos in the year 2552. Then, the number of colonists will be expanded as the resources on their new home are developed."

Angelique thanked Agwar and sat down. The light flashed before the desk of the recently elected Senator from ANZA-Pacific, and she was recognized by the Secretary.

"Mr. Cinat, please review for me how you plan to keep the number of crewmembers constant so as to conserve space and food during the transit?"

"The original crew will consist of a variety of people selected to encompass all the skills necessary to fly the starship and to set up a colony. Although the common language will be Universal English, we will allow dialects and are also aiming for diversity in terms of gender, age, race, cultural background, spiritual beliefs, etc. The only restriction will be that family groups that are selected cannot have any children older than five years. In addition, females who want children can only conceive if they are in the 30–35 age group at the time of launch. The original group of children will compose a cohort of people of roughly the same age, differing by no more than five years or so. When each child in this cohort reaches 30 years of age, he or she will be declared "Citizen" and will be permitted to have his or her own children. In this way, four cohorts will result: Learners of ages up to 29, whose role will be to grow up and learn a trade needed by the starship and future colony; Citizens of ages 30–59, whose role will be to have children in the first five years of citizenship and to perform the bulk of the routine mission tasks; Elders of ages 60–89, whose role will be to govern the starship and teach the Learners a trade or profession; and Grandelders of ages 90 and above, who will function as statesmen and consultants and help with childcare. An Elder Council will be set up to govern, in consultation with the ship's Captain. This council will mandate the number of birth licenses given out to Citizens every 30 years to take into account factors affecting the crewmember population, such as natural deaths, unnatural deaths due to accidents or illnesses, low numbers of breeding females, etc. In addition, a pool of frozen fertilized eggs and embryos will be available for laboratory birthing in case our numbers fall below quota."

"How will you prevent medical problems due to recessive gene overload?" asked the Southeast Asia Senator.

"The people involved in sexual pairing will all have genetic testing, and so will any embryos that are a product of their union. Fetuses with severe genetic handicaps will be aborted. In addition, family groups and people with similar genetic profiles initially will be placed in one of two clans: Alpha or Beta. When a child from one clan becomes a Citizen, he or she will only be able to pair sexually with someone from the other clan, and so on for future generations. In this way, we hope to keep the gene pools as diverse as possible for as long as possible, given the limits in our population numbers. After we reach Protos, the pairing restrictions will be relaxed, since the population will be encouraged to grow and since we will have an entire planet to work with in terms of space, food, and other resources. This population control system is spelled out in the Protos Mandate document, which all of the crewmembers have read and agreed to in writing."

"Which clan will the coupling pairs and their children be assigned to?"

"Unless there is some genetic reason favoring one clan or another for a given pair, they will have a choice to both become Alpha or Beta. If they can't decide, or if the numbers in each clan become dramatically different, then a lottery system will be instituted."

The light blinked on in front of Wu Cheng's desk. *Here comes trouble*, Agwar thought.

"Mr. Cinat," began the China Senator, "you mentioned earlier that you thought the S-A procedure was adequate for a century-long mission. But no one has been in suspended animation this long, or, I dare say, much longer than a few years. It would be tragic if something happened to the 40 Sleepers: death, stroke, brain dementia, and so on. Shouldn't you test your S-A procedure on Earth for decades before chancing it on a starship? I am sure the mission could be postponed during this time without undue hardship. Postponement would also allow us to recover a bit from the taxation drain that building *Protos 1* has caused. In China, we certainly could use the time to deal with more pressing issues, such as the Taiwan independence question. In fact, my government is interested in seeing if a compromise couldn't be worked out."

Nice tactic, Agwar thought. *Delay Protos, use the money saved to pay for internal Chinese issues, and offer up Taiwan as an olive branch.*

Before he could respond, the light came on in front of the Mars Senator's desk.

Oh, brother, the two of them are coming at me now, Agwar thought.

Johann stood up but looked at Wu Cheng rather than Agwar.

"I fail to see the connection between Protos and Taiwan," he said. "There is no reason that both issues couldn't be addressed separately. *Protos 1* is nearly finished and will be on its way early next year, and with the launch the budgetary drain will shrink dramatically. Then we can address other issues, like Taiwan, or terraforming Mars.

"But I think we should wait a bit longer to be sure ..."

"The Senate has been discussing this mission for 26 years," Johann interrupted the China Senator, "and I think that now is the time for action. The polls support interstellar travel, and people have been willing to spend the money to get the job done. Let's let Mr. Cinat and the Space Alliance finish the work they started and wish them good luck."

There was a murmur of agreement, and Wu Cheng sat down, with an angry glance at the Mars Senator.

Agwar noticed the puzzled look on Angelique's face, which mirrored his own feelings.

That was a surprise, he thought. *I never believed we would get support from Johann, even cynical support.*

After a few more questions from the floor concerning technical issues involving the starship, the Senate Secretary rose and asked if anyone else had a comment. When there was no response, he proposed that there be a vote to accept Agwar's written progress report, which had been sent to the Senators in advance of the meeting. This proposal was made and seconded, and a majority voted to accept it. To no surprise, Wu Cheng voted 'nay', but Johann notably voted in the affirmative. The Secretary turned to Agwar, thanked him, and said he was excused.

As Agwar collected his notes, his eyes caught those of Johann, who gave him a supercilious smile, then looked down.

8. The Captain

As the shuttle from Earth neared Callisto in preparation for its landing at Valhalla City, James Robinson glanced out the porthole. Even after more than 20 years of flying in space, the 45-year-old Captain never tired of looking at the beauty of the cosmos. But this time his gaze was greeted by a behemoth that floated majestically in the distance amidst a gaggle of orbit-stabilizing gyrotugs and supply ships. Now, in late November, the construction of *Protos 1* was completed.

The starship was indeed an impressive sight. Its four kilometer-long central core was like a silver straw stuck into the narrow end of a massive funnel, the ramscoop. This consisted of a web of thin superconducting wires that created a magnetic field to attract hydrogen ions produced from interstellar hydrogen by a forward pointing laser as the starship zipped through space. Just behind the ramscoop on the outside of the starship were the six encircling spheres that stored helium-3, deuterium, and a start-up and emergency hydrogen supply. Aft of these, and radiating out and circling the middle of the core, was a radiation-shielded, two-kilometer wide habitation wheel that housed the crew and provided artificial gravity through its rotation. Each of the four spokes that connected this habitation wheel with the central core contained supplies and a launch pad for a chemically-powered space shuttle that would be used to ferry colonists from orbit down to Protos. Rimming the rear end of the starship and pointing forward at a 45 degree angle was a series of projections that housed rockets to be used to decelerate the ship as it neared its destination.

All that's left to do is to finish supplying the ship with helium-3 from the Saturn mines, he thought, *upload us all from the Training Center, and then we'll be on our way. The ship will be mine to command!*

His reverie was broken by a buzz from his wrist compuphone. A holographic image emerged of his 39-year-old wife Melanie. He gazed at her green eyes and black hair, still enraptured by her beauty even after their 15 years of marriage.

"Hello darlin'," he said.

"Hello sweetheart. How is your flight going?"

"Just great. We will be down in about 45 minutes. I'll hop into a robotaxi and see you at the Training Center after I clear security.

"Terrific. It will be nice see you after your two months on Earth. I heard from the news reports that things went well."

James winked at her and smiled. "I just wowed them," he joked.

He paused, rubbing his short brown beard with his fingers, then continued.

"Actually, we had a few protests in New York City about having too few Buddhists in the crew, and the Houston District press complained that there were not enough U.S. State crewmembers to warrant the taxes going to the mission, but all in all people seemed supportive. The public relations folks did a good job planning the farewell parades and dinners."

"I'm glad it worked out. We need to keep people interested."

"How are the kids?"

"Doing well. Mollie is enjoying her Level 4 primary school year, and Charlie has started to walk. You will see a big change in him."

"I'll bet. It'll be great to see you all. Now that my travels are finished, we can be together and really get ready for the launch."

"Yes, the excitement down here is growing. As I was packing some of my medical diagnostic equipment for *Protos 1*, John Chandler came by to tell me that last month's psychological tests showed an upsurge in anticipatory anxiety and even some hyperactivity in many of the crewmembers. Of course, most of the anxiety came from the Sleepers as they were being oriented to the S-A protocol."

"I don't blame them. Most people wouldn't want to be so helpless for so long."

"But at least they'll be around to see Protos, if all goes well."

He was silent a moment as he reflected on her last statement, then spoke.

"I'm sure things will go well. I know that you and the engineers will be monitoring them and the S-A cryopods very carefully."

"That's for sure, although there won't be a lot of vital signs for me to look at when they are in S-A," she laughed. "Anyway, we must be optimistic."

"Yes, indeed. Well, the landing lights just went on, so I will sign off for now and see you in about two hours."

"OK, James. Love you!"

"Love you too," he said, and he deactivated his compuphone. He was glad to have had a chance to speak with his wife in real time, not in the delayed time that earlier conversations had experienced due to the distance between Earth and Callisto.

In reflecting on the conversation, James admitted to himself that he wasn't a big fan of suspended animation. He preferred a more natural approach to space missions. Ever since his wild "whiz kid" days as a handsomely muscular, skirt-chasing, rocket test pilot, and his more subdued marital days as a Space Alliance shuttle pilot, he had visited all of the colonized planets and moons in the Solar System. He was a natural to command the Protos mission, and it was convenient that his wife was an excellent space surgeon and his kids were both under six years of age.

James felt fortunate that he would have his family on board with him. It was hard for him to leave his parents and brother behind, and Melanie cried for several days after she said goodbye to her widowed mother and sisters. But not everyone could go on the mission. There were only 14 families on board, who had a total of 22 children. This meant that the remaining women in the 30–35 year range would have to partner up and give birth to 42 more children in the next few years, according to their agreement with the Protos Mandate. This was doable, given the number of women in that age category. During this period, all of the children and their parents would be assigned by the Elder Council to either the Alpha or Beta clans for later inter-clan pairing, taking into account gender, culture, and genetic diversity.

Both of his children would be in the same clan, since they came from the same family. They also would be among the oldest in the Learner Echelon, since they were born prior to launch. In fact, there was a chance that one or both of them would live long enough to see Protos. In Mollie's case, she could say that she remembered the Earth, since she spent the first five years of her life in San Francisco. That would give her something to talk about in her old age!

His thoughts turned to his arrival at the Training Center. He still had much to do in preparing the ship for the launch, still scheduled for February 1st. He would need to meet with the senior officers responsible for piloting and navigating *Protos 1*, as well as the engineers responsible for the maintenance of the fusion reactor and ion engines. He also would need to coordinate with Mission Control near the Center. Their activities would be very important during the first few months, but as the distance increased and the communication delays became insurmountable, he and his officers would be completely autonomous and responsible for the flight.

He looked out the porthole to see *Protos 1* disappear behind the edge of the ever enlarging Callisto as the shuttle descended toward the Training Center and what promised to be a busy end of the year.

9. Mike

Mike O'Malley looked out the window as the fusion-powered transfer ship arrived from Titan, where it had been loaded with its cargo. He noted the attached drums containing the helium-3 that had been collected weeks earlier from the robotic mines in Saturn's atmosphere. Although it was always tricky for the automated transfer tugs to avoid hitting the mines' suspension balloons as they approached and grappled the filled drums, it was much easier than trying to contend with the mines around massive Jupiter, with its high gravity and violent storms.

Mike maneuvered his small inspection shuttle over to the transfer ship. He would need to clear the shipment before the helium-3 could be offloaded into the storage tanks of *Protos 1*. Everything looked good to his eye, but he liked to examine the heat signatures of the drums as well. Although not mandated by protocol, as the Fuel Supply Officer for the Protos mission, Mike was very thorough in his inspection.

As he flew around the drums, he noted that Drum 3 was registering an anomalous heat source.

That's odd, he thought, *it looks fine from here.*

Mike put his shuttle in automatic docking mode, maneuvered his helmet over his short flaming red hair, locked it to his space suit, and exited the pilot seat. Leaving the ship via the airlock, he jetpacked over to the drum. His portable heat sensor registered positive in an area that outlined a flat, nearly imperceptible seal. He pulled out the knife-laser from his utility belt and carefully maneuvered it around the seam of the seal. Removing it from the drum, he saw a cavity that contained a small box with a panel of lit up numbers and some thin cables that led down to a large box underneath it.

He put his scan on the area and radioed an image back to the Control Room on *Protos 1*.

"There's something unusual imbedded in fuel Drum 3," he said. "Please have security take a look at it."

"Roger that," came the response.

After several minutes, someone replied.

"Mike, this is John Garcia. That is indeed a suspicious piece of cargo. The image is a bit blurry. Can you read what's on the panel?"

"Yes," Mike responded to the Security Chief, "It appears to be some kind of display. From left to right, it reads: '2444/12/14/0935'. Wait, the last digit just increased by one."

"It seems to be a chronometer. If the first four numbers are the year, then it could be a date and time: December 14, 0936 hours SSU, the time it is now."

"Yes, that makes sense. But this is not a standard drum component. Why is it there?"

"Mike, please scan the area with your radiation analyzer."

"OK, will do." He took the analyzer from his belt and moved it over the large box."

"Wow, this baby's hot," he said. "Its radioactivity profile indicates plutonium."

"Damn! We may have a bomb to deal with. Can the timer be deactivated?"

"Not easily. It's welded in place by three brackets, so it will be hard to get to the cables."

"All right, get yourself and your ship out of there!"

"No, wait, let me try something."

He turned on his knife-laser and began heating up the seam around one of the brackets. Slowly, the seam melted and the bracket separated slightly. He did the same to the other two brackets, then gently lifted up the timer.

"John, the timer is free. There are four thin cables going from it into the box below. Maybe I can disconnect them."

"Better not chance it. Come back to base and we'll decide what to do."

"But we don't know for sure when this thing will activate. It might trigger an explosion any time. It's too close to *Protos 1* and could damage it if it explodes."

"We can maneuver it away. Get out of there"

"No, wait, give me a minute."

Mike studied the cables. Three were red and went directly into the lower box, the fourth was black and entered at an angle into a slightly protruding area on the face of the box.

I wonder if that protrusion is a battery pack that provides energy for the timer, he thought. *If so, and if I cut the black cable, I may be able to inactivate the chronometer.*

"Mike, what are you doing?" John asked.

"I think I've found the energy source for the timer. If I cut the cable providing the energy, I can shut it down."

"But maybe cutting it will activate the bomb, or maybe it's a live cable going to the bomb that is disguised to look like something else—a decoy. Don't chance it."

"Wait a minute, I can do it."

Carefully, Mike cut into the black cable with his knife-laser.

****

In the *Protos 1* Control Room, the blinding flash obliterated the stars. Before the assembled group could screen their vision, the automatic filters were activated, dimming the light. Then the starship shook.

"A bomb" exclaimed Captain James Robinson, who had been on board for nearly three weeks. "My God!"

Next to him, John Garcia screamed into the com.

"Mike, Mike…are you there? Why did you cut it? Mike…"

"Mike's gone," said the Captain, as he switched to internal ship communication.

"Engineering, damage report."

A strident voice came back to him.

"Captain, this is Fuel Engineer Wong. It looks like there's been damage to the Storage Tank 2 entry portal. We have to suspend further loading until we can repair it. Otherwise, the ship sensors are all showing green lights."

"Get right on the repairs," the Captain ordered

He turned to his short, muscular Security Chief.

"Activate Full Emergency Protocol. And send a Red Flag alert back to the Space Alliance. We need to find out who's behind all this."

"Yes, sir!" he responded.

The Captain started to leave, then paused.

"Mike's wife is Colleen O'Malley, our Chief Computer Technician, correct?"

"Yes, Captain," John replied, "and they have a young son."

"This sort of tragedy can have a traumatic effect on a family. Better notify the ship's psychologist. I'll go see Colleen along with the chaplain."

10. Reactions

The holophone rang. The Leader activated it, and an image of a dark, heavy-set bearded man appeared.

"I have bad news," the man announced. "My inside man said they found the bomb right after the transfer ship arrived from Titan. It exploded before entering *Protos 1*. There was damage to a storage tank, but it's fixable. The end result is that we failed to seriously damage the starship. My friends on Titan are worried that they will be suspected now that the bomb has been linked to the fuel shipment, and they are going undercover. I await your orders."

Damn, the Leader thought. *The plan should have worked. How did they detect the bomb? It was small and sealed in the drum.*

He wasn't sure what to do and how to respond to his fellow conspirator. Maybe the best thing he could do was to wait and see what happened next.

****

Agwar mulled over the situation.

Who would have done it? he thought. *The project has many detractors on Earth fuming about the costs or the crew selection: too many scientists, not enough Muslims, underrepresentation of people from Asia. Many of these groups have a terrorist wing. The Solar System Police are pouring over possible suspects on Titan, since the transport ship came from there. Could China have been involved? They certainly would stand to gain by a delay or cancellation of the mission. So would Johann and his terraformists. He's been oddly supportive in our Senate meetings. Were he and his constituents somehow involved?*

Agwar got up from his desk, went over to the foodbot, and asked it for a cup of coffee and a tortecookie. Within ten seconds, both appeared in the delivery chute. He took them back to his desk.

Am I being paranoid? Could Senators do such things? There are a lot of other groups opposed to this mission.

He took a sip from his cup and a bite from his sweet. He swallowed and immediately made a call to his contact in the SSP, Inspector Gerard.

****

Angelique called her daughter, who was about to leave for a brief vacation on Phobos before coming home for the holidays. Her holoimage appeared above the phone.

"Hello Audrey, I suppose you've heard the news about *Protos 1*.

"Yes, mom, I have. What a lucky near miss. I can't believe someone would have tried such a thing. There are already hundreds of people working in and near the ship getting it ready for launch, and they all would have been killed by the blast. Too bad about the one man who tried to deactivate the bomb—I guess he'll be receiving a Solar System Medal of Honor."

"Some people are passionately against this mission and see terrorism as the only way to block it."

"Yes, I guess you're right."

"Well … will this change your plans?"

"What do you mean?"

"Do you still want to risk going on the mission? There might be future sabotage."

"Of course, mom. We can't give in to terrorists. Plus, I think the SSP will find out who is responsible and take care of the situation before another attempt is made."

"I can tell you right now that there are no obvious leads, except that perhaps someone involved with the helium-3 mining activities in Saturnmoon had the best opportunity to place the bomb. But with the mission scheduled to leave in just over a month and a half …"

"Mom, I'm going! I'm too far into it. I've got to trust that the mission will proceed as planned."

Giving in to her daughter's determination, Angelique said: "OK–I get it. I guess we'll see you after your vacation."

"Yes, mom," Audrey said, calming down. "And … give my best to dad. I've got to go, now. Try not to worry."

Seeing her daughter's image sign off, Angelique thought how difficult that would be.

****

"I'm glad Colleen and her son will be in therapy," the Captain said, pacing the room. "Losing a husband and father in this way is tough."

He turned toward John Garcia. "So, we have inspected every nook and cranny of the ship, and you didn't find anything suspicious?"

"No, everything is clean."

"Good. Harlan Wong has informed me that repairs are under way on the storage tank and that we will be able to load the rest of the helium-3 by the end of next week. The only other ships we're scheduled to receive after that are those delivering supplies and the rest of the crewmembers. Also, we can expect the occasional diplomat stopping by, garnering publicity by a visit to the starship. We must be suspicious and vigilant about all of these vehicles."

"I agree. We have developed a protocol for inspecting incoming ships by stopping and searching them when they are a thousand kilometers away. Then, they will be escorted to our docking bays by our own shuttles to make sure that they don't pick up any cargo after the inspection."

"Good. Everyone must be suspect, even diplomatic ships. We don't know who from which state might have been involved with the bomb attack last week. We seem to have detractors everywhere."

"Understood." John rose, saluted, and left the room.

James continued pacing. *We must find the terrorists,* he thought. *That is the only way we can assure the safety of this mission.*

****

At SSP Headquarters in Luna City, Inspector Jean Gerard rubbed his eyes as he activated the next holovideo. By punching in the physical characteristics of people thought to be suspects into the central holocomputer, a series of 1283 scans from cameras located throughout the city were produced that captured the potential suspects in a variety of locations over the past few weeks. This particular holovideo was taken on September 30 from a camera used to monitor the robosex activities outside of a place called the Lunatic Lounge. It showed a hooded figure walking down the street, glancing at the roboboys and robogirls, then entering the lounge. He played the holovideo back in time and noted that there were a number of people heading for the bar who were dressed much better than the typical customers of the sexbots in the area.

That's curious, Jean thought.

He went back to the image of the hooded figure and enlarged it. For a moment, the hood slipped aside, revealing the figure's face. The physical characteristics clearly identified him.

Well, look at that, he said to himself. *We have a real lead here.*

11. Confrontation

The January 2445 3-S Senate Meeting promised to be especially memorable. It was the final go/no go vote, where the Senate could approve or delay the

Protos 1 launch now that the final report had been written and circulated and most of the preparations had been completed.

"Order in the chamber," announced the Senate Secretary. "The monthly Senatorial meeting of the Solar System States is now in session. We will dispense with announcements and get right to the major item on the agenda, which is the approval of the launch of *Protos 1*. We again have with us Mr. Agwar Cinat, the General Secretary of the Space Alliance, who has requested to make the opening comments."

Agwar stood up.

"Thank you, Mr. Secretary. As all of you know by now, someone attempted to destroy or seriously damage *Protos 1* by planting a small nuclear device in a drum carrying helium-3 fuel to the starship. Luckily, the bomb was discovered by the Fuel Supply Officer, Mike O'Malley, who died trying to defuse it. It blew up, but far away from the starship. The only damage was to a storage tank, and this has since been repaired. Had the plot succeeded, the mission would have been delayed, and numerous people in the area would have died."

He paused to look at his notes, then looked up.

"I am pleased to announce that through the efforts of the SSP, the culprits have been identified."

A general buzz spread throughout the room as he continued.

"We have discovered a plot to damage *Protos 1* that was concocted by a number of people who would stand to benefit from the cancellation or delay of the Protos mission. Identified through facial print analysis obtained from a camera located near the meeting site of the conspirators, several people have been arrested who have admitted under truth serum interrogation to being involved with this plot. They have identified a number of additional collaborators, including people involved with helium-3 shipping on Titan and a technician working on board *Protos 1*. These people are being arrested as I speak. Sadly, the leader of the plot is one among you and is in this room."

Agwar paused as the chamber became deathly quiet. He turned to face one of the senators, whose violet-colored eyes glared back at him.

"Senator Schmidt, you are one of the conspirators and shall be arrested."

As members of the SSP entered the chamber and headed toward him, Johann Schmidt stood up.

"How dare you accuse me," he said. "What proof do you have?"

"We have holovideo images of you entering an establishment in Luna City minutes after some of the people who have confessed to the bombing likewise entered, and you were seen leaving said establishment shortly after they left. Furthermore, several of the suspects implicated you under interrogation as the leader of the bombing attempt."

"But you can't arrest me. I am a Senator from the State of Mars. I am representing the will of my constituents. I want my lawyer. This is an outrage!"

"The outrage was that you and your collaborators resorted to terrorism under the guise of your duties to your constituents, and that several of your group stood to gain financially from this activity."

"You are lying. You can't arrest me ..."

"But we are," said Agwar.

The SSP reached the Mars Senator, handcuffed him, and led him away screaming. When he was gone, the Secretary banged his gavel to silence the turmoil in the room. The light went on before Angelique's desk.

"The Senator from the State of Western Europe is recognized."

"Thank you, Mr. Secretary. I think we are all shocked at the revelation from Mr. Cinat and the disgraceful exit of Senator Schmidt. I am sure that he will receive a fair trial. But we must return to the business at hand. As Chair of the Senate Deep Space Committee, I am mindful that *Protos 1* is on schedule for a February 1 launch and that we must give our endorsement to this departure date. Mr. Cinat, is everything ready?

"Yes, Senator Moran. The fuel reserves are full, the supplies are nearly all on board, and the crewmembers are assembling. Pending approval from this body, the 40 Sleepers will be placed in suspended animation, and the procedure for compete fusion powerup will be implemented. We are ready to go!"

"Thank you, Mr. Cinat. Are there any questions?"

The light went on before Wu Cheng's desk. He stood up after being recognized.

"Senator Moran, Mr. Cinat. As you know, my Emperor and the State of China have been opposed to this mission, primarily because of its expense and diversion of tax funds away from pressing needs on Earth and elsewhere in the Solar System. Despite our viewpoint, we are glad that the people who resorted to terrorism to stop the mission were apprehended and that no lives were lost. China does not condone terrorism, no matter how much we might agree with the terrorist point of view. Since we believe that the Protos mission should be terminated, we shall vote to stop the launch. However, we also recognize that we are likely in the minority in having this viewpoint, and should the mission be approved, we wish the crewmembers good luck and hope for their success."

He sat down, and the Secretary asked for additional comments. When there were none, he called for a formal vote. As the roll call proceeded, Agwar reflected on the graciousness of the Chinese Senator and thought that perhaps the terrorism event might succeed in pulling all the people of the Solar System States together in support of the mission. The vote to launch was approved, with only a handful of negative votes. The mission launch was on.

12. Sleepers

Audrey boarded *Protos 1* on January 8, 2445 and was directed to her quarters. She was given one hour to store her belongings, then she proceeded to the meeting room to join the other 39 Sleepers. She sadly thought of leaving her family behind and never seeing them again, a notion that made her feel guilty. But as an only child of high-achieving parents, she needed to follow their example, and she could only do this by separating from them and being her own person. Her career was very important to her, and the opportunity this mission gave her was the chance of a lifetime.

She entered the room with some trepidation, but she relaxed a bit when she noted that people already there were casually mingling and talking as if they were at a party. Food and beverages had been placed on tables against the wall to her right, and she headed over in that direction.

"Pretty scary, isn't it?"

She looked to her right into the soft gray eyes of a handsome young man with brown hair, who like her was probably in his early 30s.

"My name is Phil Jackson, from New York City. I'm a geologist at NYU."

"Hi, I'm Audrey Moran, an astrobiologist based on Space Station 3."

"Pleased to meet you. I guess we are two of the cargo."

She smiled. "Yes, I guess so. But at least we'll be around to see our destination."

"Yes, they'll need an astrobiologist and geologist on Protos. So … what made you volunteer for the mission?"

"For me, the chance to study life in a distant planetary system. Except for Earth, the life forms in our Solar System are quite primitive, and I hope to encounter something more advanced around Epsilon Eridani. How about you?"

"Pretty much the same thing. I've been to all the major planets and moons in our system and am interested in seeing the geology somewhere else. It'll be hard to leave my family behind, however. The holidays were kind of a bust."

"Who did you see?"

"My parents and brother and his wife."

"I also saw my parents. I'm an only child. The goodbye was especially hard on my mother—she always wanted grandchildren."

"I can understand that. Too bad they couldn't go with you."

"Yes, the lucky ones get to take family members with them if the spouse had a needed skill, but we single folk are on our own. I guess for the awake crewmembers there will be pressure to pair up and start producing the next generation."

"That's the idea. We all signed the Protos Mandate, and that's the expectation. We Sleepers are lucky, however. We won't be bound by the in-transit coupling policies since we we'll be in S-A during the trip, and apparently the pairing rules will be invalidated after landfall."

"Hopefully so. But then the pressure will be for everyone to have a bunch of children in order to populate the colony and expand all over Protos."

"Well, fulfilling that pressure won't be a bad thing."

They both laughed. Audrey liked Phil. There was a certain ease about him and a similarity in their viewpoints that she found refreshing. Plus, they were about to undertake a unique experience, at the end of which they would find themselves in a select minority: people on Protos who were born in the Solar System and had recollections of Earth and the start of the great adventure.

Just then, three people came into the room, and everyone quieted down. The newcomers went over to the podium and introduced themselves as the orientation team: Leif Johansson, the Chief Engineer; Melanie Robinson, the Chief Medical Officer and wife of the Captain; and John Chandler, the Chief Psychologist. Audrey watched as the blond-haired, blue-eyed engineer walked up to the podium.

"Hello, everyone," he said. "I trust that you are all set up in your rooms, along with your bags of personal effects. I can assure you that you won't need them during suspended animation."

His remark was greeted with nervous laughter. He continued.

"Since some of you have joined the Sleeper group after the original orientation took place a year ago, we will briefly review our procedures and allow some time for questions. Over the next few days, you will receive a thorough physical and psychological examination. Then we will have a drawing to determine the schedule for when each of you will appear at your assigned cryopod for the S-A procedure. You will be bathed, shaved, evacuated of urine and feces, and receive an indwelling venous catheter. Dr. Robinson will then give you an IV sedative followed by a special cell-friendly S-A fluid that will distribute throughout your body to bathe your tissues. This contains chemicals that have been isolated from hibernating mammals. Following this will be the replacement of your blood with an isotonic energy-rich fluid that contains cryoprotectants to guard against cell damage from freezing. Actually, the water in your cells will not technically be frozen but will be supercooled to a kind of glass-like state where cellular molecular motion and metabolism ceases until you are revived. When your cryopod is sealed shut, you will be immersed in liquid nitrogen and kept this way until we reach Epsilon Eridani. Any questions?"

"Yes," said a middle aged man to her right, "I understand that the energy fluid will help sustain us as we go into and out of suspended animation, correct?"

"That's right," responded the Chief Engineer, "You will receive nutrition and oxygen to support your metabolic processes via nanobots that are dissolved in the fluid and attach to your cells. They will sustain you as your metabolism slows down to zero. These nanobots will reactivate as you are being revived when we reach the vicinity of Protos. They will keep you alive until you are cardiostimulated and begin to breathe and eat on your own in a natural manner."

"The nanobots will dissolve as you reach normal body temperature, leaving behind essentially isotonic saline that becomes part of your body's water content," Dr. Robinson added.

"What is the current record for surviving the S-A procedure?" Phil asked.

There was an anticipatory murmur in the room as John Chandler stood up. Audrey surveyed the short brown-haired psychologist with his penetrating brown eyes. She marveled at his age, perhaps no more than his late 20s. He must really have been a wunderkind to make the cut for the mission. But then, having a young psychologist would mean that he would be around for a long time to counsel several generations of crewmembers during the trip.

"It has been tested extensively in human laboratory trials for periods of up to three years," he responded. "I was involved with many of these tests, and there were no untoward physical, cognitive, or psychological sequelae of the procedure other than temporary confusion upon awakening and a memory loss of events just preceding the S-A induction. It was very safe for the subjects—just like going to sleep and waking up, except there were no dreams reported."

"Thank you, Dr. Chandler. Are there any more questions?" asked the engineer. When no one responded, he thanked everyone and invited them to continue socializing.

As she turned to get more wine, Audrey thought to herself that the issues raised in the questioning had been covered in their orientation lectures at the Protos Training Center on Valhalla. She thought that they reflected the anxiety they all felt at being in S-A for such a long time. She took a big sip of her wine and began the process of getting to know the other Sleepers, people who, a century hence, would become her friends on a new world.

****

Over the next few days of orientation, Audrey and Phil had a number of discussions about their feelings concerning the S-A procedure and found themselves growing very close. When it came time for the cryopod drawing, she picked number 7 and he number 32, which meant that she was scheduled for S-A in three days.

"Let's make the best of our time," Phil suggested after the drawing. "How about we dress up and go dancing."

Since the habitation wheel was being tested as it revolved around the central core, they had gravity, which made the dancing idea possible. She liked the plan. Audrey went to her room, unsealed her packaged clothes bag, and took out her shiny faux platinum synthetic minidress and her fancy gold shoes. After dressing, she combed her hair and put on her makeup. In honor of the occasion, she applied a trendy eyelash shimmer that subtly cycled through a palette of colors attuned to her metallic dress and shoes—from twilight blue through deep purple, then to silver, and finally gold.

Glancing at herself in the mirror, she smiled approvingly and went out. She met Phil in the hallway. He had put on a bronze metallic jumpsuit and a stylish beret.

"You look very handsome," she said.

"And you quite beautiful, he responded.

The two of them went to the ballroom and activated the roboband to play some dance music. They were alone but found themselves quite entertained with each other. After eating some snacks and drinking wine obtained from the foodbot, they danced for awhile, then decided to go back to Audrey's room. She ordered gin and tonic from the foodbot located on the wall just below the holoscreen that was transmitting a live image of Jupiter. The two of them sat down in the tight quarters, she on her bed and Phil on her desk chair.

"To us," he said, clinking her glass with his.

"Yes, to our adventure. May it be a success."

After they drank a bit, Audrey became thoughtful.

"What's wrong?" Phil asked.

"I guess I'm a little nervous about the procedure. I wonder what it will be like being in S-A under machine control. It sounds a little like being dead."

"Well, supposedly everything stops, there are no vital signs, no dreams, nothing. But unlike death, your body is preserved, and when you wake up, it's like you just fell asleep for an instant."

"I know the theory, but it still seems a bit scary."

"Well, to be honest, I feel the same way. But I just tell myself to be positive and that at the other end, things will be pretty exciting. Plus, I will have you to wake up to."

Audrey smiled, feeling herself blushing. Phil slid over to her and put one hand around her shoulder. He gave her a gentle kiss, then another more passionate one. She felt his other hand sliding up her thigh, and a general warmness came over her. Although part of her worried that things were becoming a bit complicated, she went along with her feelings as he laid her down and began taking off his jumpsuit. They made glorious love.

****

Audrey and Phil tried to spend as much time as possible together over the next three days until she was scheduled for S-A induction. He accompanied her to the chamber and stood by as Dr. Robinson and her medical team prepared her for the procedure. After receiving her indwelling venous catheter, she entered her cryopod. It was time for a last goodbye.

"Well, kiddo, you're on your way to S-A land," he said. "I'll be joining you soon."

"Yes, Phil. I should tell you that whatever happens, I will treasure the time that we were together. You mean a lot to me."

"I feel the same way. But the way I look at it, we will have years together after we wake up, and a brand new planet to explore."

He gave her a gentle kiss, then waved as he moved away. The fluid began flowing into her vein through the catheter, and she was soon fast asleep.

13. Preparations

James Robinson reviewed the launch procedures with his staff. The plan was for the starship to accelerate at a speed that would be the equivalent to a 1-g force, which would provide Earth-like gravity for the crew. The acceleration would continue until the ship reached its stable cruising velocity of 10 % light speed. At this time, the main engines would be shut off, and the habitation wheel would be set in rotation around the central core to provide gravity due to centrifugal force. This would necessitate a new orientation for people inside, with "down" shifting 90 degrees from the rear to the outside rim of the habitation wheel.

"The protocol asks us to adjust the furniture and equipment to accommodate the new orientation during the transition period between engine shutdown and wheel rotation, when we will be in microgravity. It gives us three days to do this. I assume that this time frame will be possible." He turned toward Leif Johansson.

The engineer looked up.

"Yes sir, there should be no problem. Since most of the equipment has been designed to provide displays and controls that can be used from a variety of angles, we only need to move a few things around. The same goes for the orientation adjustment when we stop the wheel during the 1-g deceleration to Protos."

"Good. Maybe we can make the orientation change a special event. The children on board will enjoy playing in the microgravity. We could have a celebration, like they did a few centuries ago when people crossed the Earth's International Date Line on cruise ships."

Everyone laughed. James then turned to Harlan Wong.

"And the stored fuel?"

"All the tanks are full, and the internal nozzles are clear. We anticipate using a bit of the stored hydrogen to get us started until the ramscoop begins bringing in hydrogen from the interstellar medium."

"Hopefully there will be enough hydrogen along our course to move us along and help us replenish our supply."

"We think so. That's the prediction based on the data from the solar sail probe that went to Epsilon Eridani."

"Are the Sleepers tucked away?"

"Yes," replied Melanie Robinson. "My medical team worked hard to safely keep on schedule, and all 40 of the Sleepers are in suspended animation, with no problems indicated."

"Give my thanks to your team," he replied. "I heard today during my walk-through that most of the food and water supplies are on-board, and the hydroponic system is functioning normally and is already producing food. So, it looks like we are on schedule for the launch date!"

14. Ad Astra

The next three days were busy ones. The final supplies were loaded, practice drills were conducted, and launch preparations were finalized. The fusion reactor had earlier been activated and was successfully receiving its helium-3 and deuterium fuel, and pumps were ready to begin flowing the stored hydrogen that would be energized and used for thrust until interstellar hydrogen could be collected by the ramscoop.

The hab wheel rotation was stopped, during which time the interior was reconfigured for the impending acceleration. A microgravity party took place the night before launch, which was scheduled for noon the next day.

At 1145 hours SSU time on February 1, 2445, the crewmembers were at their stations and ready to go. In the Control Room, Captain Robinson activated the starship's intercom.

"Well, everyone, the big moment is nearly upon us. We have trained and waited for years for this event, and now we are about to go. Leif, are all systems ready?"

"Yes, Captain, everything looks good. Fuel Engineer Wong reports excellent hydrogen flow, and the instruments indicate that the reactor is at acceleration level."

"John, are all security and support shuttles away?"

"Yes, sir, we are all by ourselves."

"Excellent. Indira, do we have clearance from Mission Control on Callisto?"

The Communications Officer looked up after a moment of listening to her earpiece. "Yes we do. Captain, Mission Control has a message for everyone. Shall I put it on the intercom?"

"Please do."

After a brief crackle, a voice boomed throughout the starship. "*Protos 1*, we are forwarding a message sent to us earlier by Mr. Agwar Cinat, General Secretary of the Space Alliance."

After a pause, the image and voice of Agwar appeared on all the holoscreens in the ship.

"To the crew of *Protos 1*. You are about to engage on one of the most historic journeys in the history of mankind. On behalf of the Space Alliance and the Solar System Senate and all its states, we wish you a safe journey. I am personally proud of each and every one of you crewmembers on board and all of the support personnel who worked to make this mission possible. Your dedication and bravery are exemplary. This heroic voyage will carry the seeds of mankind's future and will be remembered for all of recorded time."

After a pause, the launch coordinator appeared.

"We at Mission Control echo these sentiments and wish you all Godspeed. You are cleared for launch."

The Captain responded: "Thanks to all of you as well for your support and assistance. We will soon be out of range of practical communication, but we have appreciated all that you have done for us and will dedicate this mission to you, the Space Alliance, and the Senate. Farewell."

After Indira turned off the intercom, James continued: "Kiyoshi, are we ready to activate the launch protocol?"

"Yes, Captain."

"Then proceed."

The Pilot went through the checklist, activating switches and reading off instrument dials that indicated the appropriate responses. As the time ticked down to noon, there was a slightly perceptible sense of movement.

"We have begun our acceleration, Captain."

"Excellent. We are off!"

The ship gradually began to move out of orbit. The side thrusters maneuvered it in the direction of Epsilon Eridani. Man's first journey to the stars had begun.

II.   TRANSIT

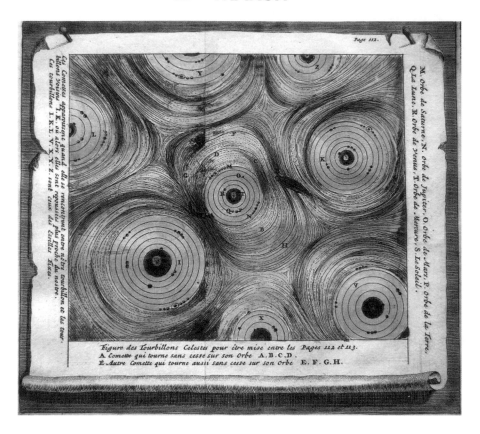

Copper engraving from Nicolas Bion's *L'Usage des Globes Celestes et Ter-restres…*, which was published in Amsterdam and bound into the 1700 edition of Nicolas Sanson's *Description de tout l' Univers en Plusieurs Cartes*. This image depicts a number of stars surrounded by swirling vortices containing their retinue of planets, according to the cosmological system of Descartes. Courtesy of the Nick and Carolynn Kanas Collection; and *Solar System Maps: From Antiquity to the Space Age,* Nick Kanas, Springer/Praxis, 2014.

15. Prologue II

Humanity's first starship streaked through interstellar space, giant in size by human standards, but dwarfed by the immensity of the cosmos. Having negotiated the Sun's planets, the Kuiper Belt, and the Oort cloud without damage, the way was clear. As the mission continued, the crew experienced the isolation of being alone and on its own, especially since the lengthening audiovisual transmission times prohibited real time communication between *Protos 1* and the Social System.

During the acceleration phase, the starship encountered an area of low hydrogen, but it was able to use some of its reserves to maintain the pace. As it approached its cruising velocity of .10c, the fuel flow was stopped and the operation of the fusion reactor was cut back to normal life support energy maintenance. The giant ramscoop continued to accumulate interstellar hydrogen until the reserves were replenished.

With the entry into cruise mode, the hab wheel rotation was again started and the crewmembers reoriented the interior appropriately as 1-*g* was established by centrifugal force. Since the giant ship was coasting its way to its destination, only a skeleton crew was used to monitor direction and look for unexpected interstellar bodies that might pose a hazard. In a sense, this was unnecessary, since some of the robotics were designed to automatically find and disintegrate any foreign intruders in the flight path via high intensity laser beams. Other robotics monitored the life support systems and the basic equipment on the ship. Nevertheless, the crew felt more comfortable providing human eyes to monitor the ship's status at all times.

During the first few years of the mission, babies were born and joined the Learner Echelon, along with those children who were part of the original crew. The adults of the original crew were assigned to the Citizen and Elder Echelons, according to age. The Alpha and Beta clans were established. They lived in two segregated areas of the ship, and they took on identifying color markings on their clothes, green and blue for the Alphas and Betas, respectively. But at work, there were no clan distinctions.

Some 30 years into the mission, everyone began to move up in Echelon status: Learners became Citizens and started to produce new Learners, Citizens became Elders and assumed their duties, and Elders became Grandelders. The cycle repeated again 30 years later as the starship passed the halfway mark in its journey to Epsilon Eridani.

In time, other than the Sleepers, only a minority of the original crewmembers were alive who directly remembered the launch, the reasons for the mission, and what it was like in the Solar System. For the majority of the crewmembers, notions of the Earth and the other home planets and moons were dependent on what they learned in history classes. The starship itself was their world, all they knew of existence. The Protos Mandate was a document connecting the past with the present, and giving directions for the future. But not everyone on board thought that it was relevant.

16. Caught

Ethan Johansson paused, glancing over his shoulder to verify that no one was following him. He continued down the empty hallway that traversed through the center of the hab wheel. Ahead and behind, this Main Corridor seemed to go on forever, without a hint of the slight upward curvature of the giant rotating wheel. As he left the green-trimmed doors of the Alpha District, he passed under a hanging time monitor at one of the four junctures where the wheel intersected with a spoke, each containing supplies and a shuttle vehicle. The date display showed '2523/01/20', and the time flashed '0350'. *Right on schedule,* he thought.

He moved quickly forward. He passed two inebriated late night revelers who were enraptured with themselves and merely gave him a nod. When he arrived in Forestland five minutes later, he looked for Sarah and felt his stomach tighten when he didn't see her. The only activity was the movement of tree leaves and ferns in the fan-driven air. He wondered if she had been discovered leaving her sleep pod.

"Ethan," came a whisper from the shadow of a willow tree to his left. Sarah Sandal emerged, wearing a short green-trimmed minidress. They embraced. He caressed her silky black hair and felt the warmth of her body.

"Did you have any problems with the sleep pod sensors?" he asked.

"No. Like we planned, I programmed them to show an empty hallway, even when I passed by. Elder Dukas taught us well in computer class, but he'd be surprised to know how we're using his training."

He chuckled, then kissed her. She anxiously pushed him away while glancing around for interlopers.

"Ethan, I'm scared. What will happen if we're caught?"

"We'll tell them they have no right to restrict us. We should be able to see each other."

"But the Protos Mandate …."

"Was established when the ship was launched decades ago. By now, our Echelon should be able to set up its own rules."

Before Sarah could comment, the lights suddenly brightened to the sound of an alarm. Ethan grabbed her hand. "Damn! Let's get out of here."

They ran down the Main Corridor toward the hydroponic farms. A shot sounded and they tumbled down, enveloped in a sticky fibrous net.

"Not so fast!" the security guard barked as he came toward them brandishing a restraint rifle.

As he sprayed them with the release fluid that dissolved their fibrous prison, Ethan held up his head defiantly and responded: "Leave us alone. We're not hurting anyone."

"You two know the rules against being out after curfew, especially for male and female Learners from the same clan. We saw you two kissing on the hidden security camera monitors."

Ethan's mind raced, seeking a way out. "Citizen Ortiz, your son is a classmate of ours. We're fellow clansmen. Let us go back to our sleep pods. Nothing bad happened here."

"No can do. If it got out that I'd let you go, I'd be in trouble myself."

He nudged them forward in the direction of Security.

17. Punishment

Elder Christopher Kormos rose stiffly from his bed after the morning bell awakened him from a deep sleep. He had worked into the night composing his monthly progress report for Earth. Transmitted at the speed of light, it would take some seven and a half years to reach Space Alliance Headquarters due to the great distance involved.

Until we figure out a way of moving faster than the speed of light, we all have to be patient in this era of interstellar travel, he thought. *The Alliance has already begun planning follow-up missions to other stars, so I guess our reports are useful, however long it takes them to reach Earth.*

He stretched his lean muscles and massaged his fashionably shaved head as he turned toward the sink. Looking into the mirror, he saw a new wrinkle on his face.

The price I pay for chairing those cantankerous Elder Council meetings, he thought. *I was trained to be a pilot, but now all I do is supervise a bunch of old fools arguing over trivia! Resource-willing, someone else will be picked at the next council election.*

Trimming his graying beard, he rehearsed the highlights of the 78th Launch Day speech he would deliver on February 1: *cruising velocity remains fixed at 10 % light-speed … estimated date of arrival at Epsilon Eridani is July 4, 2552 … reserve hydrogen tanks full and adequate for course corrections and planetfall deceleration … fusion power ample for life support and hab wheel rotation … Learner, Citizen, and Elder Echelon populations stable at 65, 64, and 62 people each, with 49 Grandelders over 90 ….*

The phone alert buzzed and was followed by the saccharin computer-generated voice he hated. "Good morning Elder Kormos. You have a call from Security."

Hiding his annoyance at the interruption, he activated the holophone and saw Citizen Manuel Ortiz gazing back at him.

"I'm sorry to bother you, sir, but I found Ethan and Sarah from Alpha Clan in Forestland earlier this morning. They were alone and were kissing. I have them here awaiting your recommended course of action."

"So I take it they were in violation of the Learner curfew as well?"

"Yes sir, they were. It was before 0700 hours."

"Very well. Thank you. Keep them there. I'll be over in 30 minutes."

He donned his blue-trimmed jumpsuit and selected his breakfast from the foodbot. He was irritated that his favorite cereal was unavailable due to the monthly rationing. As he waited for the food to arrive, he thought about Ethan and Sarah.

Some of our Learners are becoming a handful. They want to do things their own way and disregard the Protos Mandate. "Too restricting," they say. But we can't afford to loosen up, given our limited space and resources. I must make an example of the two of them.

He ate quickly, left his room, and began walking to the Business District, the quarter of the hab wheel that housed Security, the Control Room, the Mass/Energy Converter, the Ballroom, and all of the other work areas. Beyond, in clockwise succession around the wheel, were the other quarters: the Alpha District, Forestland and the hydroponic farms and ranches, and finally the Beta District where he lived.

Ten minutes later, he arrived at Security. Guarded by Citizen Ortiz, the two 17-year-olds sat glumly handcuffed to their respective chairs. He noticed that with his blond hair and blue eyes, Ethan looked a lot like his father, who came from a succession of engineers going back to Leif Johansson, who was Ethan's great grandfather. Sarah, by contrast, had green eyes and naturally black hair with dyed orange streaks running through it, the current retro-fashion for young Learner girls. With the Founding Generation ship's captain and the original chief physician as her great grandparents, Sarah also had famous people in her family tree. Nevertheless, both she and Ethan had to follow the rules like everyone else.

"Well, what do you two have to say for yourselves?"

Sarah spoke first. "Elder Kormos, we were just looking for a quiet place to talk. There's always someone around in the day time."

"You were violating curfew. And I am told you were being … intimate."

"We just wanted to be together," she pleaded. "What's the harm?"

Kormos reflected a moment on the Mandate reproduction section that dealt with limiting the spread of unhealthy genes and producing approximately equal-sized Echelons whose exact numbers took into account unexpected deaths and factors that might strain the limited resources of the starship.

"You know the pairing rules: 'Male-female pairings can only take place between Citizens from opposite clans, and all babies must be licensed and born to mothers aged 30–35.'

Ethan glared at him. "But we weren't exactly having sex. Also, most of our Citizenship will take place on Protos, where there will be ample space and resources. Why should we be forced to pair and have babies with someone we don't love in the other clan?"

Kormos bristled. Although over a century ago, the unmanned Epsilon Eridani probe had shown the fifth planet of the star to be rocky, in the habitable zone, and to have the temperature, atmosphere, and water to sustain human life, no one knew for sure if people could thrive and reproduce there.

"Because we will be in transit when you become Citizens at age 30. The Protos Mandate will still be in force and must be obeyed."

"Sarah and I love each other. By the stars, the rules need to be changed for our Echelon," demanded Ethan.

"I'm surprised at you," he told the youth. "Your father is an engineer, and you yourself are in the engineering apprentice track. You of all people should know the fixed resources on board this vessel and how important it is to manage resources by controlling our population."

"But times have changed," the boy argued. "The Elders need to be more reasonable and relax the rules for our Echelon. We must be allowed to …"

"That's enough!" roared Kormos, annoyed by the youth's persistent arguments. "You violated the law. You will each spend 40 hours doing starship maintenance duty. And you must never again be together unsupervised."

Kormos looked over at Ortiz, who was recording the verdict. When the security guard finished and looked up, the Elder stormed out of the room.

18. Microgravity Basketball

The next day Ethan continued to seethe over the judgment, even as he was dressing for the yearly inter-clan M-grav Basketball Game after dinner. It was to be played in the special microgravity gym and dance area located in the part of the central core that was the hub of the hab wheel. The baskets were placed five meters high, necessitating high acrobatic movements to score points. Ethan was tall, strong, and athletically-inclined. He loved playing the game, especially when it was an inter-clan match, and at tip-off the events of the previous day faded from his thoughts.

The Beta team led his Alpha squad by a score of 140–137. On the floor, fans from all the Echelons were belted to their seats and cheering for their favorites. Ethan huddled with his team during a time out as the coach discussed their attack. The plan was to score, then stop Rufus O'Malley, who already had scored 56 points. Tall and strong, with flaming red hair, Rufus was a terrific athlete but sociopathic and a bully when he didn't get his way. A year ago, he broke another boy's arm in a fight over a seat at a crowded dance party, and until recently he had been on restricted activities for ration violations. However, he was allowed to play basketball for the blue squad.

As the time buzzer announced the resumption of the game, Ethan glanced at the clock: 48 seconds left. He put the ball in play by throwing it to his friend Vadim Zubkov, who with long light brown hair floating wildly in the microgravity pivoted to his left. The rules dictated that no more than three steps could be taken on the court. Then, the person with the ball had to stop or lift his toes against the tongue plate in his shoes to deactivate the sole magnets from the metal floor. This would allow him to push off and soar into the air, either to go for a basket or pass the ball.

Vadim took two steps, then went up into a high arc. Ethan followed a step later by pushing off toward the basket. At the zenith of his jump, Vadim pushed the ball in the direction of Ethan's trajectory. Ethan caught it just beyond the reach of his defender and stuffed it into the basket. The sensors in the net recorded the score with a bell ring, and the scoreboard read 140–139.

At the next inbound, Ethan had a feeling that the ball would be thrown to Rufus, so he began running toward him as it was passed. They both jumped in the air. Ethan managed to deflect it, and it careened off of Rufus' face before hitting the out of bounds cushion. Several players laughed as Rufus rubbed his nose and glared at Ethan. It was Alpha ball, with just 18 seconds to play.

During the time out, the coach told Vadim to try for the game-winning basket. He ran down the court and jumped as the ball was passed. A blue-shirted player deflected it down into Ethan's hands. He noticed Rufus moving in for the block

I've got to get around him, Ethan thought.

He took two steps, pushed off the floor in a forward motion, and began to somersault. Rufus jumped as well. He threw out a large fist to block Ethan, but he missed the rotating target. As Ethan jammed the ball through the basket, the net bell sounded just before the time buzzer. The final score: Alpha 141, Beta 140.

As Ethan pushed himself down from the rim, he glanced beyond the snarling Rufus. Amidst the celebrating fans, he spotted Sarah cheering for the victory. She smiled and discretely pursed her lips in a kissing motion. He smiled back as his teammates surrounded him.

19. School

The next day, Ethan got up early for school. He shared a room with Vadim, who was in the adjacent bathroom. He quickly made his bed and dressed. The room was small, just large enough for their two beds with trundle and overhead storage for clothes and personal possessions, and their two desks with overhead light and personal holocomputer. The walls were decorated with posters of Earth and the Solar System, and there was one faux window screen that electronically displayed stars to give the illusion of looking out into space.

Ethan and Vadim lived with other Learners in their Echelon. They had been brought up by Grandelder parent surrogates and Elder teachers and lived in a common area. They saw their birth parents on weekends and special occasions. Along with the few children whose parents chose to marry for life and raise their children themselves, with the help of Grandelders and robonannies, they attended a common primary school until reaching their teenage years. Then they moved in with a same-sex roommate and attended four levels of high school, receiving a general education. In Level 3, they all took vocational tests, which determined where they would do their specialty apprenticeship training after the following year. Each Learner had some choice in the matter based on interest and aptitude, but the available slots were determined by the Elder Council in conjunction with starship need, and the final placements were determined by the high school principal and counselors. Now in Level 4, Ethan had been placed in the engineering track, and Vadim in the medicine track.

After both of them finished dressing, they left for their tutorials. On the way, they discussed the basketball game.

"That was quite a finish," Vadim said. "You made a great shot over Rufus. He was really steaming afterward."

"Well, it was nothing personal. I know that he's a real hot-head and has gotten himself into trouble over the years."

"That's for sure," Vadim said. "From picking on people to stealing things in the Beta District—he's always had a bad reputation. And all those fights at school … My dad says that it runs in his family. Both his father and grandfather were always getting into trouble. I would watch out for him, Ethan. He really knows how to hold a grudge. He probably thinks that you wanted to show him up during the game, and he's probably planning some kind of retribution."

"I try not to pay him any attention, it only encourages him," Ethan responded.

"By the stars, I hope you're right. But be careful."

All school classes were conducted in the adjacent Business District of the hab wheel, so they only had a twenty minute walk. Afterward, Ethan would attend his engineering tutorial with the Chief of Engineering, and Vadim would go to his medical tutorial in the hospital section. They were already beginning their transition to specialty work. After graduation, each would get a new roommate who was in the same apprentice track, and they would only see each other socially from then on. Ethan would miss Vadim, his friend since childhood.

"Are you seeing Sarah later on?" Vadim asked.

"Yes, this evening when she finishes her hydroponic tutorial. She's always loved botany and helping things grow."

"Where are you meeting?"

"In the Spiritual Room."

"Well, be careful. You don't want to be caught again."

They walked a moment in silence, then Vadim spoke.

"Do you think things will ever change about pairing?"

"I don't know. The rules are really antiquated and rigid. There was a lot of anxiety when the mission was launched about our small population and the chances of encouraging harmful recessive gene expression in future generations due to genetic drift. That's probably why the rules ended up being so authoritarian. But in reality, we have complete genetic analyses and periodic updates through life to look for mutations, so any pairing can be examined for problems, and some genetic match difficulties now can be fixed due to recent advances."

"You got that right. I'm thinking of doing my medical thesis on this issue. Our docs are doing some really exciting new work in genetic modification, work the Founding Generation knew nothing about."

"The same with engineering advances. You know, I've been assigned to clean up the M/E Converter Room as punishment for seeing Sarah in Forestland, but the cleaningbots can easily be programmed to do what I'm doing manually. I think cleanup is an outmoded punishment—everyone knows it's make-work in the service of the legal system."

"Well, we don't have room for a prison on board," Vadim said as he slid open the tutorial room door, "so manual work has to serve as a major punishment for criminal activity. You know, they used to make convicts break up rocks on Earth as punishment for crimes. Can you believe it?"

Ethan laughed. "I guess it could be worse for me, so I shouldn't complain."

They walked in and took their respective seats for the tutorial.

The roboteacher gracefully walked into the room. She looked just like a woman of 30 years, except her skin was perfectly smooth, without a blemish or wrinkle. She was dressed in the latest starship style: retro 1960s. Her synthetic hair was gracefully combed in an attractive flip, and she wore a trim metallic minidress, as did most of the young women on board. The young men wore one-piece jumpsuits, with a low front that exposed their upper chests and in many cases a metal medallion. The clothes glistened in the gold, silver, and copper hues of the recyclable synthetic metals of which they were made and were surprisingly comfortable. All were trimmed in either green or blue, reflecting the Alpha or Beta clan membership of the wearer. Older citizens had similarly trimmed clothes, but they were more modest in style and usually made of recyclable durapaper in pastel hues.

The roboteacher, whose name was Ms. Delta, was programmed with many more facts about the history of the Solar System than any human counterpart could possibly possess. She was networked to the large holoscreen in the room, which was activated as she discussed certain points. Today's tutorial was on the history of planetary expeditions during the twenty-first and twenty-second Centuries. The fifteen Level 4 Learners, who were in the last year of general education before beginning their specialty apprenticeships, had all studied the lesson the night before on their holocomputers and during sleep learning, so they knew the material pretty well.

Ms. Delta began with a survey of the main manned missions in sequential order: the Moon bases, Mars, Jupiter and its moons, Venus, Saturn and its moons, Uranus, Neptune, and finally Mercury. She reviewed the major commercial enterprises that followed, including the mining stations and the colonies that took place on planetary, moon, or asteroidal surfaces and in space stations revolving around them. She tied these into the global warming and overpopulation pressures on Earth. She finished with the preparations for the Kuiper Belt and Oort Cloud missions in the twenty-third Century that led up to the star exploration probes, which would be the topic for the next tutorial. All in all, Ethan thought it was an excellent review.

During discussion time, one of the students asked: "Ms. Delta, why couldn't we simply stay in the Solar System and just terraform planets and build more space stations to accommodate all the people?"

"It was all a matter of economics," she responded. "Terraforming is expensive, and with the possible exception of Mars, very complicated. You have to remove the dangerous atmospheric gases and adjust the pressure and temperature with giant vacuum machines and solar light collectors. Then you have to add carbon dioxide to stimulate plant growth. If done properly, you can later introduce simple animal life. It takes time and money to arrive at the right conditions to support mammalian and human life. Space stations are cheaper and easier to produce, but they have limited space and resources to handle the population needs."

She paused a moment to consult a different file stored in the memory banks in her head.

"Interstellar missions like this one are also very expensive. But they are a better investment for the future. Finding a suitable planet is easier and cheaper than terraforming a hostile planet or moon. In addition, the starship costs will likely decrease as we improve the design and reliability of our propulsion technology, space ship architecture, and suspended animation capabilities."

Ethan shivered as he thought about the 40 Sleepers: *They wouldn't feel too happy to hear the last comment!*

"What about warp drive?" Vadim asked.

Again, a pause from the teacher to find the correct file.

"That will be the topic of a future tutorial. But let me just say now that when we embarked on our expedition, Solar System scientists were working on ways to gravitationally alter space-time to produce such short cuts to the stars. In addition, anti-Einsteinian scientists at Neptune State University were claiming that they had mathematical proof that you could go faster than the speed of light using a tachyon drive."

"But we left some 78 years ago," Ethan retorted. "What about now?"

"Well, our starship physicists do not agree with these calculations. They believe that light still holds the speed limit. "

Everyone laughed. Ethan thought about Ms. Delta's ability to make a joke. Was that excellent programming, or was she capable of symbolic thought, or even human-like consciousness?

"In addition," Vadim added, if the Neptune State scientists were correct, their space ship would have reached us by now. I don't see them anywhere!"

"That's assuming they would want to contact us. Maybe they would be so far ahead of us in science and technology that we would be too old-fashioned for them. They may prefer to go their merry way to Epsilon Eridani and not pay attention to us primitives, preferring instead to welcome us after we ploddingly arrive at Protos."

There was more laughter, but some pause as the students considered the implications of what she said. Before more questions could be asked, the ship time bell rang, signaling the end of the hour and of the tutorial session.

20. Caught Again

That night, Sarah sat in one of the pews of the Spiritual Room, the only place on board where privacy was sacrosanct and cameras were forbidden. Since she had reserved the room for the hour, no one else was present.

I shouldn't be here, she thought. *I really want to see Ethan, but the risks we are taking in meeting …*

Her thoughts shifted to the recent fight she had had with her parents, with whom she had visited the past weekend. Her father thought she was throwing away her life, daring further censure from the Elders, all for a boy from her own clan. Her mother, who was the child of Mollie Robinson, the daughter of the original starship Captain and the Chief Medical Officer, was worried about appearances and how their family would be shunned by the other citizens. In this closed community, word traveled fast, and everyone knew about

Kormos' verdict. Sarah felt like an outcast and a disappointment to her parents.

The door opened and Ethan tiptoed in, locking it behind him. He sat down next to her. After giving her a brief kiss, he confessed: "My family's not very happy about what we're doing."

"Neither is mine."

"But damn it, Sarah, the Protos Mandate is too rigid and outdated! The rules have taken away our freedoms. They need to be changed."

Sarah grabbed his arm. "Shh, calm down. Someone may hear you."

"I don't care. I'm tired of being told what to do. And we're not the only ones unhappy with things."

"What do you mean?"

"The Returnees. They never signed up for all these rules either!"

"But they are fanatics. They want to scrap the mission and go back to the Solar System."

"Yes, and they have a point. Even though their previous relatives signed away their rights to go to Protos, they never did, and they don't see any reason to continue on to a foreign planet around a foreign star."

"But their actions are pretty radical. We don't even know if the ship has the resources to turn around and spend another four decades returning to who knows what."

"Still, Sarah, our freedoms are limited, and for what? An old outdated mandate that we never agreed to."

Suddenly, there was a commotion outside. The Spiritual Room door burst open to reveal Elder Kormos, Citizen Ortiz, and two other security officers.

Red-faced, Kormos spoke: "You two continue to disobey the rules."

"How did you find us?" Ethan demanded.

"We put tracer pellets in your food and have been monitoring both of you. When the two tracers converged here, we knew you were together. You have violated my decree and the Protos Mandate again. You must now go before the Elder Council, who will decide your fate"

21. Danger

The next day, Ethan was serving his maintenance duty obligation in the M/E Converter Room. The last technician had left early, and he was mopping up the floor where some of the waste had lain before being shoveled into the converter to produce supplemental energy for the starship.

Sarah and I will probably get another 40 hours, or even more, for meeting in the Spiritual Room, he thought.

He heard the door open behind him and turned to see a blue-clothed figure entering the room. It was Rufus, sneering and red in the face.

"Well, basketball jock, nice to see you doing some real work."

"What do you want, Rufus?"

"I thought I would come and pay you back for trying to show me up by slamming the ball in my face at the game."

"It was an accident and you know it. I was just blocking the shot."

"Bullshit! I don't like being humiliated."

Rufus reached toward one of the tool shelves, grabbed a bolt pistol, and pointed it towards Ethan's foot.

"With all the equipment lying around, it's too bad you accidentally shot yourself in the leg while cleaning up."

"You bastard! Give me the pistol."

Raising it, Rufus said: "Not so fast, or the accident will be in your chest."

"Why do you want to do this? I never gave you any cause..."

"Oh, but you did, at the M-grav game. And you're such a goody-goody, talking the talk but just wanting to change things to suit your own needs. Grabbing up Sarah when she should end up with someone in Beta. Never mind the rules, it's all about you."

"None of us should be told who we can and can't be with. I want to be with her and she wants to be with me. We should have our free choice of partners. YOU should have your choice too."

"I like the pairing rules just the way they are. Maybe I want Sarah to be with me and not you! She's a real hot number. We could have a nice life together on this starship. Hell, I would just as soon keep flying in space and not worry about all the hassle to set up a new colony ... Anyway, when the Returnees finish the job, we'll be going back to Earth, so it won't matter."

"What do you mean, finish the job? What are they planning?"

"Never mind, you'll see soon."

Rufus pulled the pistol trigger, and a bolt shot out trailing a wisp of compressed air. It just missed Ethan, ricocheting into the control panel and extinguishing two panel lights.

"You fool!" said Ethan. "You could do some damage in here"

"Then maybe you should get in the way to prevent it."

He shot again, but the bolt missed the moving target and smashed into the wall. Ethan scurried behind a desk, under which he saw a red security button. He pressed it.

"Come out, you coward."

As Rufus came around, Ethan pushed the desk chair and it hit his adversary in the leg. As he recoiled, Ethan jumped at him, grabbing his gun arm. The two of them battled together for what seemed to Ethan like an eternity until he managed to push Rufus away and scurry through an opening in the bulkhead, which was hit with a clang by a speeding bolt that just missed his head. On the other side, he saw a side door and tried to open it, but it was locked from behind.

"Now I got you," Rufus said as he came in, raising his gun.

Suddenly, the main door to the Converter Room opened, and Rufus became immersed in a fibrous net. Trailed by two security guards, Citizen Ortiz entered the room, lowered his restraint rifle, and looked at Ethan.

"What's going on in here?" he said.

"Rufus tried to shoot me with the bolt pistol. He was sore about the basketball game."

"Take him away," Ortiz directed his men. Then he turned to Ethan. "You're very lucky that we were nearby when the alarm went off and got here in time. Come with me ... we need you to file a report."

As Ortiz started to leave, Ethan held him back.

"You know," he told the security guard, "Rufus said something about the Returnees finishing a job, something they'll be doing soon."

"What do you mean?"

"I don't know for sure, but I think they have a plan to send the ship back to Earth."

Ortiz was thoughtful.

"That group is a big problem. They are always yacking about something, always threatening. But they back down in the end. We don't know of any plans, but they'll need to go through us if they're thinking of doing something funny. Come on, I'll take your report and tell my superiors what you said."

The two of them walked out toward Security.

22. Consultation

After dinner, Elder Kormos received a phone call from Ortiz.

"Yes Manuel, what's up?"

"We just learned from our truth serum interrogation of Rufus O'Malley that there's some sort of action being planned by the Returnees to send us back to Earth. He doesn't know the plan exactly, but it bears watching this group more closely. Citizen Ratchel especially."

Ratchel, thought the Elder. *He's been very passionate lately in his speeches before the Elder Council about the rights of the Citizens to make up their own minds about returning to Earth versus continuing on to Protos. Now we have Ethan and Sarah raising a ruckus about the pairing restrictions of the Protos Mandate. I think it's time to have a talk with Chandler.*

"Thank you. I agree with the increased surveillance. Keep me informed."

He hung up and made a call to the psychologist, who at age 106 was one of the oldest people on board. A white-haired, wrinkled visage with still penetrating brown eyes greeted him on the holoimage.

"Hello Christopher, nice of you to call," the image said.

"Hello John. You look well."

The old man laughed. "I'm still vertical, so I can't complain. Resource-willing, the anti-aging treatments will continue to keep me going. Yesterday I was told that I have the mind and body of a 90-year-old! But I have a few aches and pains, so I'm sure I'll be ready to enter the M/E Converter along with the other waste products when I reach the mandatory death age in four years. At least I'll be adding a little power to the ship rather than taking away consumables."

He paused thoughtfully, then continued.

"Anyway, you didn't call to hear an old man grumble. What can I do for you?"

"I'm picking up some resistance among the crewmembers to our mission. The Returnees seem to be ratcheting up their activities, and some of the Learners are questioning the pairing restrictions of the Protos Mandate. Do you have a sense of what might be going on?"

The psychologist thought a moment. "Well, we are roughly three quarters of the way to our goal, with two complete Echelons and part of a third being born in space. None of these people signed the original covenant for the mission, and they weren't around to see the problems on Earth and in the Solar System that prompted it. So, they haven't completely bought into the expedition psychologically. All they see are the rules and restrictions. In addition, many probably won't be around when we reach Epsilon Eridani, so for them the goals of the mission are not their goals. I sometimes feel the same way, although I do remember the climate changes produced by global warming and the overpopulation on Earth, so these memories keep me going."

"What do you think we should do about it?"

"Maybe you could call a town hall meeting to discuss the mission and show holovideos of Earth at the time we left."

"I'm not sure that would help much. It might make people more interested in our home planet and Solar System, and it might spur more anger at the restrictions imposed by the Mandate."

"Tell me, Christopher, which problem do you fear the most, the Returnees or the complaints of the Learners?"

"The Returnees. They are better organized and threaten everything we're doing. The Learners just want to have more freedom to pick a partner of their choice from either clan. This is a freedom they will have on Protos anyway."

"Well, why not give it to them?"

"But the Mandate ..."

"Has served us well for 78 years. Now, with just 29 years to go before landfall, maybe we could loosen up a little. I remember that there were disagreements at the time of launch about all the rules. With genetic counseling availability, some thought the idea of clans were too restrictive. Others saw this as an additional safeguard against people accumulating too many recessive genes by marrying someone related to them. But we have successfully kept the clans separate for nearly eight decades, so maybe we could relax things a bit now."

"There are some on the Elder Council who might object."

The psychologist laughed. "Well, Christopher, you can be very persuasive. Plus, maybe they will be convinced if they see that there is popular support for this notion. Is there?"

"I don't know. Two of our 17-year-old Learners, Ethan Johannson and Sarah Sandal, both from Alpha clan, have been caught kissing and being together in violation of the Learner curfew, and they have declared a desire to pair up when they reach Citizenship. But I'm not sure how many others hold this view."

"What are you doing with them?"

"Both will be going on trial in a couple of days since they have violated the Mandate twice, which means there will be an open forum that all crewmembers can attend."

"This is a good thing. You will have a large group in attendance, and you can get a sense of how much support there is for loosening the rules."

"And I suppose that one way or the other, we can then move on and focus our attention on the Returnees."

"Exactly. The open forum should prove very enlightening."

"Thank you, John. Your advice is always helpful."

"My pleasure. By the way, how is your speech going for the celebration?"

"Very well, thank you. It's always good to take stock of our progress on the Launch Day anniversary."

"Yes it is. I plan to be there as well, health permitting."

"Good. If so, feel free to make comments about what it was like in the early days of the mission, especially at launch. People always like to hear what you say."

"From the horse's mouth, I guess."

They both laughed at the quaintness of this ancient expression. As he hung up, Christopher thought how lucky they were to have a few of the original crewmembers still alive to help put things in perspective.

23. Trial

The Ballroom was abuzz. Most of the starship inhabitants were present to view the spectacle. Not one, but two crimes were on the agenda, both involving Learners: a crime of violence starring Rufus, and a crime of passion starring Ethan and Sarah. What a show!

The nine Elder Councilors sat at a semi-circular table at the front of the room, with Kormos in the middle. After discussing a number of routine issues related to the mission, Rufus was tried. He was convicted and sentenced to a three-month cooling off period in confinement in Security, followed by five years on restricted activities. But the real sentence was the scorn he would receive from the starship populace, with nowhere else to escape to.

Ethan and Sarah then were called forth and took two seats at the focal point of the semi-circle in front of the mass of crewmembers in the Ballroom. Kormos spoke.

"Learner Ethan and Learner Sarah, you are brought before this Council for violating the Protos Mandate twice by having private and intimate intra-clan contact. You also violated my proscription against seeing each other unchaperoned. The Council will be meeting shortly to decide on an appropriate punishment, but first you have the right to make a statement. Have you anything to say?"

Ethan looked at Sarah, then responded: "Learned Council, Sarah and I love each other and want to be married when we become Citizens. We realize that this is in violation of the Protos Mandate. But we'll spend most of our adult years on Protos, where the rules will be different. The Mandate needs to be changed."

There was a murmur in the crowd, particularly among the Learners.

One of the other Councilors responded: "We have not yet reached our destination, and the Mandate is still in force and will be until after we land in 29 years. It must be obeyed …"

Ethan was pleased when Sarah interrupted with the same revolutionary words that were on his mind.

"The Learners were born under this system but never agreed to the Protos Mandate, like the First Generation. It needs to be modified to account for the future realities of our Echelon."

The crowd murmur became louder. A Learner in the crowd stood up and yelled: "Sarah and Ethan are right. We're being held back by old-fashioned principles that we didn't choose."

A number of Learners and Citizens stood up and applauded. Kormos banged his gavel and they sat down.

Another Citizen then rose and tried to speak about tolerance, but she was drowned out by an enraged Elder yelling for the laws to be maintained.

Kormos stood up and banged his gavel. "Silence! There will be order or I will call Security."

The crowd quieted down, but Kormos was mindful that most of the comments he heard were in support of change. He continued the trial proceedings.

"How do the two of you plead?"

Ethan declared: "We are guilty of disobeying a rigid rule that has little relevance for our Echelon today."

Sarah added: "And we are guilty of being in love! Is that a crime?"

Chaos erupted, as several people in the audience stood up and began to speak. While his fellow Councilors looked at each other with dismay, Kormos again called for order. At the back of the room, he saw a familiar person who was raising his hand."

"I believe Grandelder Chandler would like to speak."

The crowd silenced as the old psychologist walked up to a microphone.

"Yes I would. Thank you, Elder Kormos."

He paused for effect, then continued.

"I was around when we launched, and there were many people who were against this mission. We even had a terrorist action against us that threatened to delay or sabotage the launch. But the majority of people on our polluted Earth and in the Solar System gave us support, because they saw the problems stemming from overpopulation and believed that off-loading our numbers to exoplanets would help balance the population/resource equation. The Solar System was filling up, terraforming was too expensive, and the stars became the hope. In a similar manner, we have a population/resource problem on this ship. We need to keep our numbers in control until we land on Protos. We must therefore control the number of babies born to match the number of people dying, and those who are born must be healthy and functional. The Protos Mandate was developed to guide the *Protos 1* society in a sustainable manner. Resource-willing, we will make it to our final goal."

There were calls of "hear, hear" and "well said" from the floor.

Chandler continued.

"Nevertheless, it was never the intention of the Founding Generation to needlessly restrict people's freedoms, nor to carve the Mandate in stone. It is a guide, not a shackle. So as conditions change, we must be willing to change the Mandate to accommodate new realities. After all, every new baby receives a thorough genetic analysis, with periodic updates to look for mutations. That will likely suffice to clear new pairing couples from now until we reach landfall."

Many Learners and Citizens, and a few Elders, stood up and applauded.

Some of the Councilors looked at each among the buzz. Kormos banged his gavel again for order. "Thank you, Grandelder Chandler. We have much to consider. Any other comments?"

Hearing no further response, he rose, and the Council members walked out of the Ballroom into a side chamber to discuss the verdict.

****

During the deliberation, the general buzz in the room reflected discussions about the possible verdict and Dr. Chandler's comments. Ethan and Sarah waited and fidgeted. She then turned to him.

"What do you think will happen? Will we be punished further?"

"The worst punishment would be separation from you. I love you and want to be with you always."

"I feel the same way," she said, tears welling in her eyes. She took his hand and put her head on his shoulder while they waited.

Thirty minutes later, the chamber door opened, and the Councilors filed back into the Ballroom and took their seats. The crowd became silent. Kormos spoke.

"We have deliberated the fate of the two young people before us. Their violations of the Protos Mandate and my decree are serious matters. But we have taken into account some of the statements made by Learner Ethan and Learner Sarah. There is a case to be made for the fact that they will be spending much of their adult lives on Protos. In fact, their argument applies to all members of the Learner Echelon. The words of Grandelder Chandler also were taken into account and have important implications for the future. But we first need to deal with the matter involving the defendants, and we have reached a decision."

Many in the crowd leaned forward in anticipation.

"The Council has decided that Ethan and Sarah must devote another 40 hours to starship maintenance for their second offense, and they must not see each other unchaperoned. Furthermore, the Council will meet over the next few days to review the clauses of the Protos Mandate and to make any modifications that seem reasonable given the current situation. This will pertain as well to the pairing rules, with possible ramifications for the defendants."

Random clapping and cheering echoed in the room, particularly among the Learners.

Kormos turned toward the couple. "Do you abide by this judgment?"

Ethan was about to protest their separation from each other, but Sarah held his hand and whispered: "This will likely be only a temporary separation. Let's see what happens."

"OK," he muttered.

They both turned toward Kormos and voiced their agreement. He nodded and banged the gavel, closing the trial.

24. Grandma Mollie

Sarah knocked on the blue-trimmed door.

"Come in," came the voice from within.

Grandma Mollie was sitting before her computer. The 83-year-old dietician was reviewing algae calorie counts for her talk to the Level 4 high school students, her gray and black bun bobbing up and down as she compared figures in the columns.

Turning toward her granddaughter, she said: "Welcome dear! Please get some tea from the foodbot and have a seat.

Sarah walked to the machine past a wall covered with photos of several Solar System planets and moons, as well as pictures of her granduncle Charlie and her mother as a young girl. She took cups of tea from the 'bot for the two of them and sat down on the bed next to her grandmother.

Finishing her work, Mollie looked at Sarah with still vibrant green eyes.

"What can I do for you, dear?"

"I guess you heard about the trial yesterday?"

"Yes I did. It was covered live by the holovideo news. You and Ethan have certainly stirred things up."

"I know. Mom is really mad at me. She thinks I'm bringing disgrace to the family."

"Don't you worry about it, Sarah. The Protos Mandate is too old and rigid. John Chandler was right—it needs changing."

"But mother thinks our good name will be tarnished by some of the Elders."

"That's just too bad. You know, your great-grandparents sometimes had to take unpopular stances as well, and they now are the stuff of legends. But at the time, wow! People really were mad."

"What do you mean?"

"I remember my parents arguing strongly that no family should be broken up when the clans were established. Many people felt that we should adopt a collective mentality, where people paired simply to have children, and all babies were taken into the nursery, bottle-fed, and reared through childhood by older citizens. This would free up the parents to focus on the work of the starship. But some people thought that families should remain intact and that Citizen parents could still perform their work duties if older people helped out with childcare. There weren't any Grandelders in those days, but with the anti-aging treatments available, most people could continue to live until the mandatory death age and could provide a great service to our society by helping to raise children."

She took a sip of tea, then continued.

"My parents and most of the original family members were in favor of the 'family first' plan, whereas most new couples liked the collective idea. They thought this fulfilled the spirit of the Mandate and would allow for a smoother running social system on board the starship. This created a vigorous debate, with the traditionalists arguing that the Echelon system created enough of a disruption to family units, and the collectivists accusing the family first people of being old-fashioned and not keeping up with the new realities demanded by life on a multigenerational starship. They decided to vote on the matter, and after some rancorous debate, the collective idea won out and was established and used by the majority of new couples. But the family plan remained as an option. My parents selected it, and I was brought up by them."

She took another sip of tea.

"So our family has been involved with social issues and Mandate interpretations from the get-go. Your mother seems to have forgotten this bit of history. But then, when she paired up and joined your father in the Alpha District, they elected to go the collective route, which is the way you have been brought up."

"Wow, that's interesting about my great grandparents. What else do you remember from the launch days, grandma?"

"I was only five at the time, but I remember having to stay inside to avoid the pollution on Earth and the excitement related to the launch and leaving the Solar System. It was touch and go for a while, since we passed through an area of low hydrogen and needed to tap into our reserves. I recall my father worrying about this issue until the hydrogen densities picked up again. Then there was the S-A scare of 2462. The power went off that led to several of the cryopods, and they began to heat up. My mother and the engineers were in a real tizzy about possible damage to the Sleepers. They finally found the source of the local power outage, fixed it, and the Sleepers seemed to be OK, but we won't know for sure until all of them wake up during landfall … Then there was the black hole fiasco."

"Yes, I read about that in class."

"But what you didn't read was why we almost entered its gravity well. The year was 2476. I was 36, your mother was a year old, and your great-grandparents were Elders in their 70s. Captain Foster was in charge then, but he was relatively inexperienced, having taken over just a few years earlier during the calm cruise phase of the mission. When they detected the space-time distortion from a small black hole, some of the scientists wanted to fly closer to examine it. It would have only meant a slight turn off-course using some of our side thrusters and very little fuel from our reserves. Captain Foster agreed. But we were traveling fast at .10c, and no one had ever attempted a course diversion at that speed before, and he had no experience moving the starship. Anyway, we came too close, and *Protos 1* drifted toward the event horizon. It was only a rapid acceleration from the main thrusters that got us away from the pull."

"Wow, I heard about the close call, but not the reason."

"That's the problem with a multigenerational ship. The mission is long, and when a new Echelon takes over control, they are relatively inexperienced and haven't had to deal with all of the ramifications of their work. Our system of computer training and elder mentorship is a good one, but not infallible. Luckily, my father was still around to help Captain Foster out of his jam."

"You certainly have seen a lot, grandma. And know a lot. My classmates are really looking forward to your calorie talk."

"Well, dear, I've been a dietician for over 60 years and have seen a lot of changes in our hydroponic system. Hopefully, I can teach all of you a bit about nutrition."

"I'm sure you will."

She got up to leave.

"And don't you worry about this thing with Ethan. I was lucky in finding your grandfather, who came from the Beta Clan. We had very similar values, although at the time some considered us old fashioned. When we met, he was an apprentice in the hydroponic section, where I did my first dietetic rotation. As you learned in school, he later became the director of the section and was instrumental in boosting the production of the algae farms. He also made the algae tastier through the additives he developed, and people became more inclined to include this food product in their diets. Anyway, when we paired off after becoming Citizens, we decided to sign a marriage contract with a life provision, not one of the decadal contracts. We didn't think much of raising our kids in the communal school, preferring to have them live with us until their apprenticeship. When I moved over with him to the Beta District, I made a lot of new friends, as well as keeping the old friends I had from my Alpha days. We really had a good life together until he died in that freak electrical surge seven years ago."

Her eyes teared over, but with a slight sniffle she continued.

"But getting back to your situation, there is nothing wrong with finding someone in your own clan, someone you have known for a long time. Love and common values should guide pairing decisions, not a set of abstract rules that have outlived their purpose. After all, we are well past the halfway point of our mission, and the end is in sight. Most of your life will be on Protos, not on this ship. ”

"Do you think the rules will be changed, grandma?"

"Anything can change if people want it bad enough. I'll talk to some of the Council members. I have known most of them for decades. Maybe I can persuade them to be a bit more flexible."

"Thank you, grandma."

"You're welcome, dear. Well, you best scoot along. I need to finish my algae count review and get my school presentation in order."

Sarah gave her grandmother a hug and went out the door.

25. Conspiracy

Jonathan Ratchel nervously prepared for his meeting with his fellow Returnees. It was in a small room adjacent to the Navigation Center, where he worked as a Navigator 2nd Class. The room was just large enough to house the four of them. Ben Juarez was the first to appear, having just gotten off duty in the Fusion Reactor section. Red Barbosa was next, brushing the feed from his overalls that he had collected during his work in the Farm Animal Quarters. Finally came Amir Nasser, always late despite working close by with the M/E Converter.

"OK, we can start now," Ratchel said. He was a small, swarthy man with a twitchy tick that pulled up the right corner of his mouth. "Let's do a final run-through of our plan for the takeover tomorrow during the Anniversary Celebration in the Ballroom. I checked the security schedule that was published an hour ago. There will be one guard posted in the Control Room, and I will be the only other person there. The Chief Navigator was delighted that I volunteered to cover for him so that he could go to the celebration along with the rest of the crew. Red, you come right over at 1320 hours SSU time, and when the guard opens the door, shoot him with the animal tranq pistol you smuggled out from your job, and the Control Room will be ours."

"Got it," said the huge man with reddish brown hair, still smelling of chickens and feed.

Ratchel continued.

"Ben and Amir should be in the Fusion Reactor Monitor Room by then with the bomb—Ben has the key to get in. I will then get on the intercom with our demands."

"Are you sure you can control the side thrusters so that we can turn the ship around?" asked the short, dark-haired converter tech.

"Yes, don't worry, most of the work will be done by the central computer anyway," Ratchel responded. "After we make our announcement, I'll program the ship to move into a large loop ending with a return vector to Earth. Last night I calculated the trajectory very carefully. It should only take 55 hours to complete the loop and fine-tune the return course. When this is completed, I will destroy the yaw, pitch, and roll thruster controls to keep the ship unalterably pointed toward our Solar System for the time it takes to return, and the future crew will have no choice but to decelerate again or bypass our Solar System for deep space. We'll be very old or dead by that time, but this travesty of a mission will be stopped. Red, are the supplies now in position?"

"Yes," the burly man said, rubbing the red birthmark on his neck. "I stored enough water and supplies in both the Control Room and Monitor Room cabinets to last us for several weeks until we achieve our goals."

"Good. Resource-willing, our plans will proceed smoothly."

Ben opened the bag he was carrying. A tall man with squinty brown eyes and hair, he pulled out a cylindrical device with two suction tips and a timer switch on the outside. He handed it to Ratchel.

"Here, after you push out the guard, you can seal shut the outer metal Control Room door and blow up the adjacent door control panel with this bomb. With the inner fireproof security screen manually closed and locked, no one will be able to get in."

"Excellent," the navigation officer responded. He then produced his own bag.

"Here are some laser pistols that I procured from the officer's armory. They may come in handy."

He gave one to each conspirator.

"It's really too bad we have to do this," Ben said wistfully. "We tried and tried to convince people to go back home, that it was against the laws of the universe to journey to another star. The stars are placed far away for a reason: alien life forms are not supposed to intermingle."

"Yes," replied the converter tech. "We've seen how our species contaminated the microbes on Mars when people first arrived. And then we tried to terraform the planet like we owned it. We polluted the Earth, our first home, as well. We'll do the same to other star systems. If we continue this way, we'll violate the harmony of the cosmos..."

"But no one listened to us!" Ben interjected. "They said the Returnees are just another wacko group, some sort of religious fundamentalists. Yet the fact is that we have no god other than the universe, and the laws are irreversible."

"We have to take matters into our own hands. We're doing the right thing," Ratchel added.

He put the small bomb into his backpack.

"Until tomorrow, gentlemen," he said. "The plan is afoot."

The group disbanded, with each member leaving at two-minute intervals and going back to their own rooms for the night.

26. Intruder

Ethan had mixed feelings. On the one hand, he was glad that his and Sarah's actions had stimulated the Elder Council to consider changing the Protos Mandate to allow for more flexibility in pairing relationships. On the other hand, he was despondent at their separation, even for just a little while. Plus, he was assigned to clean up in the M/E Converter room today of all days, when the whole crew would assemble in the Ballroom and he might have a chance to see Sarah, however briefly. But that was his fate, and he must learn to live with it, at least for now.

"Well, Learner Johannson, things are looking much cleaner," came the strident voice from behind him.

He turned to see his supervisor, a thin black-haired man with steel-gray eyes.

"Thank you, Citizen Nasser, I am doing my best."

"Well, I think that should be enough for today. You can leave, now."

"But I must complete eight hours every day until I fulfill my sentence."

"I'm on duty here, and I say when your shift is completed. Go. I assume you'll want to see the celebration."

"If I don't log in a complete tour of duty ..."

"I'll log you out myself at the proper time. Go along, now."

As he left the room, Ethan thought that this situation was odd. Nasser had never been friendly to him, and he was mystified as to why the tech should want to miss the celebration himself when Ethan was required to be there for several more hours.

He walked down the Main Corridor toward the Ballroom amidst a crowd of crewmembers heading in the same direction. The program was scheduled to start at 1300 hours, forty minutes from now. He kept worrying that Elder Kormos would see him and discover that he was not serving his time. He turned around and started to head back. Up ahead he saw Nasser leave the Converter Room, lock the door shut behind him, and head in the direction of the Beta District.

That's strange, he thought. *He shouldn't be leaving. And he's heading away from the Ballroom.*

Ethan followed behind Nasser, both of them going against the oncoming flow of people. When Nasser came to the wheel spoke just before the Beta District, he turned into it and proceeded toward the central core. As gravity began to decrease, he stopped at a cabinet and grabbed a pair of Velcro over-shoes and slipped them on. Ethan did the same when he reached the cabinet. This allowed both of them to walk in weightlessness on one of the special Velcro strips that proceeded down the center of each of the four walls of the corridor.

When he reached the core, Nasser turned and continued walking toward the front of the ship. He paused before the Fusion Reactor Monitor Room and slipped inside.

That door is supposed to be locked at all times, and only the technician on duty has a key to the FRM room, he thought.

Ethan came to the door, opened it slightly, saw no one close by, and slipped in. The room was very long, with monitoring dials and reactor switches lining the two side walls. At the end was a giant filtered screen that allowed the techs to look into the core itself, but the iris-like cover was constricted shut. With their backs to him inspecting a satchel were Citizens Nasser and Juarez, one of the techs. Ethan slipped behind a partition that demarcated a small internal office space.

"You had no problem bringing the package?" Nasser was saying.

"No, it was easy. I just carried it from my room along with some reactor maintenance charts. I'm sure people thought it was all part of my monitoring equipment."

"Good. Jonathan and Red will likely be in position shortly ... it's nearly 1150 hours now."

"Since you're here, I'll lock the door and we can await their call," Ben said as he turned to walk back. Ethan kneeled behind a chair in the dark office and was not seen by the reactor tech as he passed the partition. He locked the door and returned to his comrade.

"Now we wait."

A few moments later, they received a call on the station to station intercom. The caller's voice was familiar to Ethan, but he couldn't exactly place it."

"This is Citizen Ratchel calling from the Control Room. Is anyone there?"

"Yes, Jonathan, this is Ben. Amir and I are here and in position."

"Excellent. I'm here with Red. There was no problem with the guard. He's asleep outside in the hall, and the two of us are sealed in. Your door control destruction device worked very well. Probably everyone has settled in the Ballroom by now, so no one will pass by the guard. Are you ready with your special package."

"Yes," said Ben. He stroked the satchel. "The bomb is here and ready to be armed. We have the camera pointing to it and to us in case someone wants proof that we mean business."

"OK, we'll wait until the contact time and then activate the ship-wide intercom."

Ethan became anxious. *Bomb! What's going on here? I've got to warn the crew.*

He changed his wrist compuphone from voice to text mode and quietly typed a message to Vadim to notify security that he was locked in the FRM room with Citizens Juarez and Nasser and that they had a bomb. Also, that Citizen Ratchel was in the Control Room and was somehow involved. Then he waited.

27. Launch Day Anniversary

The ship chronometers read '2523/02/01/1255' as the crewmembers gathered in the Ballroom. On the stage were Captain Samuel Markman, Pilot Ward Brown, and Elder Kormos. All the spectator seats were full, with latecomers standing in the back. Some 220 people were present, nearly everyone except for a few security personnel and officers manning critical areas. There was a buzz in the air as people anticipated the celebration of the starship's most important holiday.

At the top of the hour, a recorded version of the Solar System States anthem began to play as everyone stood at attention. At its conclusion, Captain Markman took the podium. He was a tall, hefty 51-year-old man with white-dyed hair that gave him an air of formal authority. He welcomed everyone on behalf of the officers and gave a brief status report of the ship's progress. Next, Pilot Brown stood up. In contrast to his leader, he had a dark complexion and jet black curly hair. Although diminutive in stature, he was wiry and strong, and he looked much younger than his 48 years. His speech was sprinkled with jokes, providing levity to the program. Finally, Elder Kormos stood up. He gave a speech honoring the significance of the occasion, and then said he had an important announcement to make.

"Ladies and gentlemen, the Elder Council has been meeting to consider the issue of relaxing the Protos Mandate's pairing rules. This consideration was prompted by the recent trial of Learners Ethan Johansson and Sarah Sandal, both in the Beta Clan, who have expressed an interest in courting one another with the possibility of a future pairing. The Council considered two issues related to this trial and its implications for the current Learner Echelon. First, we will land on Protos in 29 years, and members of this Echelon will live most of their citizenship and elder years on the planet's surface. Second, we have made advances in genetic screening and are able to find and in some cases neutralize the effects of dangerous recessive genes in future pairing individuals. Consequently, the Council has decided to rescind the pairing rules from this day forward for the Learners, who will be the next Echelon eligible for birth licenses."

Most of the crowd erupted in cheers and applause, although there were a few whistles of protest from some of the Elders in the room.

"Of course," Kormos continued, "citizens requesting birth licenses must undergo genetic analysis in the year prior to pairing in order to rule out undesirable recessive genes or recent mutations that might affect their offspring. In addition, this amendment to the Mandate will not affect the policy pertinent to the issuance of birth licenses, which will continue to be based on the need to keep our population in line with the resources available. Finally, effective today, the sentences for both Learner Johansson and Learner Sandal will end, and they will be pardoned and able to see one another freely and without supervision."

More applause from the audience. Sarah was in attendance and ecstatic at the news. She looked around for Ethan but did not see him.

Kormos continued to make a few more announcements. Then near the end of his appearance, the ship intercom crackled, and a voice came on.

"Attention crewmembers of *Protos 1*. This is Navigator 2nd Class Jonathan Ratchel speaking from the Control Room. I am announcing that my fellow Returnees and I have taken over the ship. We are secure in the Control Room and the FRM room."

A buzz went through the crowd. The Captain and Pilot stood up in shock. Kormos glanced to his right, as the Chief of Security ran up to the stage.

The intercom voice continued: "Our intention is to turn the starship around and head back to the Solar System, where our species belongs. After doing so, we will destroy the thruster and navigational controls with a bomb and metal acid, and we will completely erase the computer navigational programs with a targeted virus. The ship will continue only on a direct course toward Earth and will be able to decelerate but not veer from its path in any other way. Our brave soldiers have enough supplies to last us until our plan is completed. Any attempt to interfere will result in our blowing up the Fusion Monitoring Room."

The intercom clicked off.

"They mean business, Elder Kormos," the Chief of Security whispered as Markman and Brown came up to the podium. "We just heard from Learner Vadim Zubkov that he received a text message from Ethan Johansson. He's apparently trapped in the FRM room with Citizens Juarez and Nasser, who have some sort of bomb."

"So this is all happening," Kormos said. "Is Learner Johansson part of the plot?"

"Apparently not. He texted that he followed Citizen Nasser into the room and is hiding in a small internal office area."

"That might be a help to us," Kormos said.

He turned to the crowd.

"Ladies and gentlemen, this is indeed a serious and unexpected situation. Please return to your duty stations. I will confer with the ship's officers and the Elder Council to plan our course of action."

He turned and directed Markman and Brown into a corner as the crowd dispersed.

"Any ideas, gentlemen?"

"Well," said the Captain, "maybe we can get Ethan to do something about the people in the FRM room. But that still leaves the Control Room."

"I have an idea," said the Pilot. "The ventilation system connects all areas of the ship and is big enough to enable a person to crawl to the vent in the Control Room. If we can distract Ratchel and whoever else is in there with him, I can try to make my way to the vent and perhaps neutralize them."

"That might be worth a try," said Markman.

Kormos looked at the Pilot.

"Do you think you should do it, or perhaps someone else? It would be very confining."

"Listen, Elder, in my apprenticeship training I had to squeeze into these shafts and remain there in the dark for an hour to show that I was not claustrophobic. I have also had to cram into a tiny shuttle cockpit during pilot refresher training. So I'm used to small spaces. I just need to study the ship's ventilation map on my computer to figure out the best route to take."

"OK," said Kormos, "you sound like our man."

The three of them exited the room, and Brown headed for his quarters in Alpha Section to study his computer. Kormos and Markman headed for their rooms in Beta.

"I need to consult privately with some of the members of the Elder Council," Kormos said turning toward Markman. "What do you think of the Returnees' plan?"

"I don't know. They seem to think that they can successfully turn the ship around toward Earth and irrevocably damage our thruster and navigational capabilities. We would have 78 years to assess the damage and figure out some way of fixing things and turning back toward Protos. But they could certainly cause a lot of damage, and they may be right in thinking that we could not undo their damage."

He thought a moment more, then continued.

"Frankly, I'm more worried about what they might do to the reactor. If they explode a bomb in the FRM room, they could breech the reactor wall and turn loose a lot of radiation, or even cause a massive explosion. I'll have to consult with the engineers about the risks."

"Maybe Ethan can do something about this."

"Maybe, but he's young, and these terrorists are determined and bold. I don't know what he can do."

"Do you have a plan for him? We could text him some directions via his friend."

"The monitoring room may have some heavy tools like a wrench, or a bolt pistol, or even a laser gun for cutting metal. I don't know. I'll consult with the fusion experts to see what might be available. We'll do what we can. We have two problem areas, and it would be very helpful if Ethan can help us resolve one of them."

28. Takeover

Ethan felt a slight pull of gravity to his right.

They must be turning the ship, he thought, steadying himself in the micro-gravity.

Just then, he felt a vibration on his wrist compuphone.

A text appeared. It said: *E ... Kormos message: laser gun in equip cabinet on L wall near work bench ... subdue terrorists ... security outside door ... V*

Subdue the terrorists! Wow, what a job, Ethan thought.

He considered what he could do. He carefully peeked through a window in the office partition. Both men stood before a partially opened satchel that exposed the bomb and the handles of two laser pistols. Halfway between him and the men was a cabinet on the wall. Maybe he could make it there before they discovered him. But was he up to it?

I guess I'll never know until I try, he thought. *And now is as good a time as any.*

He quietly maneuvered himself over to the office entrance opening, crouched down with his feet on the Velcro strip, then sprung toward the cabinet. His time playing M-grav basketball served him well, as his jump was accurate and quickly moved him to the cabinet. As Juarez and Nasser turned toward him in surprise, he jerked open the door, saw the pistol with its charge light on, and pulled it free from its energy connection.

"Don't move!" he said. "Hold up your hands and slide away from the bomb and the pistols."

"What the hell are you doing here?" demanded Juarez.

"He's just a kid—let's get him!" roared Nasser.

Juarez sprung. Ethan reflexively fired the pistol. A spot of light hit Juarez in the left shoulder. He screamed, and his movement spun him to his right where he landed near the cabinet.

"You bastard!" he wailed.

"I told you not to move," Ethan said. He pointed the gun at Nasser.

"Now, you get away from the bomb and join your buddy."

Nasser quietly floated over to his fellow conspirator.

"Give me the key to the door or I'll shoot you closer to your heart."

With his right hand, Juarez took the key from his pocket and floated it over to Ethan, who grabbed it and carefully moved over to the door. He unlocked it, and three security guards came in on Velcroed feet. They handcuffed the terrorists, despite some whimpering from Juarez as one of them roughly pulled his arms behind his back. The other guard wristphoned the successful capture back to his headquarters and grabbed the satchel with the bomb.

As they were leaving, one guard turned to Ethan.

"Nice work, son. You did us all a big favor. If you ever want to change your career path, let us know. We would be glad to apprentice you."

"Thank you, Citizen."

As Ethan walked behind them down the central core toward the junction with the four hab wheel spokes, his wrist compuphone vibrated. It was Sarah. He stopped and flipped on the voice button.

"Ethan, are you all right?"

"Yes. It has been quite an adventure."

"The starship intercom has announced the successful capture of the two terrorists in your room, and it said that you helped in the capture."

"Yeah, I had entered the room unseen by them and managed to stop them with the aid of my trusty laser pistol."

"Everyone is saying that you are a hero. I'm so proud of you."

"Well, I sure didn't expect to be one this morning when I woke up. But I guess the crisis isn't over yet. Any word about the situation in the Control Room?"

"No, nothing."

"By the stars, we will get them as well."

"Yes, I hope so. But I have some other news."

"What?"

"Before all this terrorist stuff happened, Elder Kormos announced at the celebration that the Elder Council has pardoned us and decided to relax the pairing rules. We can see each other any time we want. Isn't that fabulous!"

"Wow, that is good news. Where are you? I'll come over right away.

"I'm near my study pod in hydroponics. Why don't we meet in Forestland?"

"The location of our original crime! That would be ironic. Now it'll be the location to celebrate our freedom."

They both laughed as he continued down the corridor. He signed off as he reached the juncture. He turned toward the spoke that led to the Alpha District/Forestland boundary of the hab wheel. He felt the force of gravity steadily increasing as he neared the end of the spoke.

It will be good to see Sarah and to have a more normal relationship, he thought. *I'm glad the Council changed their ways. This will be better for all of us Learners and for our offspring.*

At the juncture, he turned and headed for Forestland.

29. Capture

Ward Brown crept within the major trunk of the ventilation system. He had entered the system through the wall in Navigation, just adjacent to the Control Room. He remembered that the branch of the system going to the Control Room was just ahead. He reached it and looked down through the ventilation cover plate. He couldn't see anything though the slats, but he could hear the men below as he quietly began unscrewing the plate. It was just large enough for him to get through. He checked his wrist compuphone. The time showed that it was 1600 hours sharp.

As earlier planned with Kormos and the Security Service, a moment later there was a buzz below. He heard a voice.

"Yeah! Ratchel here. What do you want?" it said.

The intercom echoed in the room.

"This is Elder Kormos. We have discussed the situation in the Council. We are asking you to give up and come out peacefully. I assure you that your demands will be discussed and that you will be given a fair trial."

"The time for discussion is over," said Ratchel. "I've appeared several times before the Council, and they've ignored me. It's a travesty to be going to another star. We've taken action to remedy this situation. We're turning the ship around."

"You should know that we have captured Citizens Juarez and Nasser and defused their bomb. They are in custody."

"So, that's why they haven't responded to our calls. Well, you won't get us! In fact, you must release them, or we will start to shut down some of the power to the wheel, making it uncomfortable for the crew. We'll start with the lights and ventilation to Alpha and Beta Districts and the hydroponic systems."

"We'll find ways to block you and manually take over the ship," Kormos said.

"You're bluffing. The central control is here, and we can override everything. As a navigator, I know how the systems operate. And Barbosa knows the hydroponics section. You can't fool us."

As they argued and were distracted, Brown removed the last screw from the cover plate. He sat on the right angle juncture between the main and control room ventilation pipes, kicked the plate down where it fell with a crash, and followed feet first into the room. He saw Ratchel and Barbosa standing before the intercom speaker. A large container of metal acid was on the floor nearby. They looked incredulously at him. He landed with his laser pistol pointed at them.

"Hands up!" he said.

Both reached for their own pistols. He shot Barbosa in the chest, while Ratchel dove behind the large captain's chair. Brown's shot hit the seat with a sizzle. From the side, Ratchel shot back at him. He felt a searing pain in his right ankle as he hit the ground and rolled. Another spark hit the floor just beside him. He stopped rolling and aimed at the side of the chair, where Ratchel again appeared, gun aimed at him. Both fired. A sizzle hit the wall next to his ear. His shot was more accurate, burning a hole in the navigator's right eye. He went down groaning, then became silent.

Brown looked down at his ankle. His foot had been largely severed from his leg and was attached only by some skin and tissue on the inner side and by part of the fabric of his boot. Blood was pouring out. He quickly pulled off his belt and tied it just below his knee as a tourniquet. He slid over to the control area and pulled himself up to the intercom controls.

"Brown here. I have subdued the terrorists. Both are dead, but I've been wounded in the ankle. There's a lot of blood."

"Good work," said the voice of Kormos. "We'll get you out shortly."

He noticed that the outer door controls were destroyed.

"You'll have to cut me out with a torch," he said. "You won't be able to open the door any other way. And don't delay."

After unlocking and sliding open the inner fireproof security screen, Brown sat down, feeling woozy and nauseous. He tried to keep the make-shift tourniquet tight as he waited. The blood kept oozing out, and the pain was severe. The wait seemed an eternity.

Five minutes later, a welding flame appeared through the top right corner of the door and began to move around its periphery. The bright flame seemed to be progressing at a slug's pace. Brown felt very light-headed and weak, relaxing his hold on the tourniquet, which made the blood flow more profusely. Finally, the flame path completed its circle, and the door separated from the jam. It fell inward with a loud clang. Two security guards and two medics entered. While the security personnel surveyed the scene, the medics came over to attend to Brown as Kormos and Markman entered.

"Good work, Ward," the Captain said.

"A piece of cake," Brown slurred, sweat forming on his brow. "These guys knew how to cause trouble, but they didn't know how to handle a pistol."

"Well, they knew enough to find your ankle," the Captain said, with a worried look on his face.

Brown gave a slight smile, then passed out. The medics quickly put him on a stretcher and whisked him away to the hospital section. The security guards began to remove the two bodies.

The Captain was over at the control console.

"What's the damage?" Kormos said, joining him.

He looked up with concern on his face.

"They managed to begin a broad turn," Markman said. "And it appears that they messed with the software as well." He pointed to the metal acid container. "Lucky for us, Brown got to them before they had a chance to pour some of that into the controls."

"Can you fix the software?" asked Kormos.

"I hope so. I've called Chief Navigator Alcindor to come help me. He's a good man who really knows how this navigation software works. If we can't fix this, we'll continue to turn forever, going nowhere, but resource-willing, we'll get back on course.

He paused and looked at the Chief Councilor.

"We'll have to boost security to make sure that no one else causes more mischief,"

"These four are the main Returnees that kept bugging us at the Council meetings. I don't think that there are any more of that group left. We'll just have to figure out a way of restraining Juarez and Nasser. Rufus O'Malley is serving out his time in the only seclusion room in Security, but two of our lockable storage rooms were originally planned as convertible jail space for criminals. We'll have to quickly empty them for these guys. We can't let them loose again."

Kormos left the Captain and the newly arrived Chief Navigator to do their job. As he headed back to Security to make plans for the prisoners, he worried about Brown and decided to swing by the hospital as well. This day was turning out to be quite different from the way it had started.

30. Resolution

Over the next five days, Alcindor and his staff worked on the software. The pressure was great, as the starship continued to arc uselessly in space. The evening of the fifth day, the Captain checked in with him.

"How is it going?"

Pushing the auburn hair from his brow, the Chief Navigator looked up with earnest brown eyes.

"I don't know for sure. The rebels really made a mess of the programming. They've even put in viruses that are activated when we try to make corrective changes. But I think we've taken care of everything. We may as well try it and see what happens. If the system crashes, we're back to ground zero."

"I'm sure we'll be alright," Markman said, not quite believing it himself.

As Alcindor entered the modified coordinates that would put them back on course to Protos, the Captain looked around. The other crewmembers in the Control Room were gravitating over to them, concern on their faces.

They all watched the three dimensional holoindicator that showed the red line that represented the ship's current circular course. Nothing changed.

"Let me try some manual operations to see if I can help the system along," Alcindor said.

After a few minutes more, the red line on the indicator slowly began to move toward a green line that pointed in the direction of their destination. Numbers flashed in the periphery. Alcindor smiled.

"It's working! The starship is responding and moving out of its arc and toward the correct course. We're back on track toward Protos!"

The assembled crewmembers erupted in cheers.

****

Several days later, a special town hall meeting was held in the Ballroom. The occasion was to celebrate the course correction and to award special commendation certificates for the heroes who helped quell the uprising. After some introductory comments concerning the successful end to the crisis, Kormos walked over to Brown, sitting nearby in a wheelchair with a large bandage around his reattached and surgically repaired right foot, but smiling. He gave the pilot his certificate amid enthusiastic applause. Brown responded with some characteristic jokes, but then seriously thanked the hospital staff for reviving him and saving his foot.

Kormos then walked back to the podium and motioned for Ethan to join him.

"Learner Johansson, it gives me great pleasure to award you this certificate for extraordinary service to the starship. By subduing the terrorists in the FRM room, you helped us end the crisis at the risk of your own life. It was a brave task and one that has earned our gratitude."

There was robust applause as Ethan stepped up, accepted the certificate, and made his way to the podium.

"Thank you Elder Kormos. I did what I had to do, and fortunately things worked out for the best. I'm happy to contribute."

He looked out and the crowd and saw Sarah, smiling and applauding. Ethan continued.

"I would also like to thank the Council for their action in modifying the pairing rules. It will be a wonderful thing for my Echelon as we mature toward Citizenship, giving us flexibility to find a proper mate no matter what their clan is. It will also set the stage for a nice transition as we approach and then land on Protos. I think I speak for all of us Learners in thanking you and the Elders in the Council for your flexibility."

More applause, as the meeting came to an end.

Ethan stepped down and Sarah ran up to him and gave him a kiss. They hugged for a moment, then Ethan walked over to Kormos.

"Sir, may I have a moment to speak with you in private?"

"Certainly," the Elder said, and the two of them moved to the side of the room.

"Sir, I would like to transition my career path to Security. I know that my family has produced engineers, but there are several Learners pursuing this path. I have always been athletically-inclined, and I have an interest in doing the right thing. I know that Sarah and I tested the system a bit, but we meant well, and I am impressed that the Council was flexible enough to change an outmoded rule. I can support and defend the modified Mandate, and I am ready to show it through an apprenticeship in Security Service."

Kormos thought a bit, then said: "Well, son, you know that career changes need to be approved by the Council, but we do in fact have a surplus of engineer candidates, and Security can always use bright, athletic people with proven bravery. I'll see what I can do for you at the next Council meeting."

"Thank you, sir," Ethan said. They separated, and he went over to Sarah and gave her another hug.

"Well," he said. "Elder Kormos was open to my career change. I hope it's approved."

"It should be," she countered. "How can they turn down a starship hero?"

They hugged and walked out of the room.

III. PLANETFALL

A panorama of the Martian surface taken by the Viking 1 lander after it touched down on July 20, 1976, the first spacecraft to land on Mars and successfully carry out its mission. Note the two-meter wide rock partially covered with red soil to the left of center, which was named "Big Joe." Uncovered portions were similar in color to basaltic rocks on Earth. Will other planets around distant stars look like this? Courtesy of NASA/NSSDC, with collaboration from M.A. Dale-Bannister (Washington Univ. in St. Louis); and *Solar System Maps: From Antiquity to the Space Age*, Nick Kanas, Springer/Praxis, 2014.

31. Prologue III

For 28 more years, *Protos 1* continued on its path toward Epsilon Eridani without incident. A new generation of Learners was born, as their Citizen parents took on major starship duties. The former Citizens became Elders, the Elders Grandelders, and the cycle of life continued. The Grandelders alive at the time of the Returnee Mutiny had all died, and their bodies had been disposed of in the M/E Converter. Most of the deaths occurred via the Passing Ceremony at the mandatory death age of 110, but a few were due to accidents, illnesses or voluntary early suicide from people who believed that it was better to die with dignity on one's own terms rather than at a mandated time.

The latter choice was selected by John Chandler, who at the age of 109 decided that he had had a good life and that enough was enough. His life celebration was attended by nearly all of the crewmembers, and he seemed to thoroughly enjoy the public comments and the private goodbyes. Then to tumultuous applause, he was escorted into the Passing Room by the medics. Here, he was bathed, gently messaged with scented oils, and given the passing juice. He lay down and watched holoimage recordings of Earth's few remaining forests to the sound of Beethoven. As his muscles relaxed and he gently fell asleep, his last thought was how nice it was to end a full life like this. His body was wrapped in synthetic linen and taken solemnly to the Converter Room for its final disposal.

In mid-2551, the crewmembers began to prepare for their encounter with Protos in July of the next year. The giant hab wheel's rotation was slowed down, then stopped, as the habitable areas were reconfigured and oriented to a new source of gravity. The deceleration rockets at the rear of the core were activated and the ship began to decrease its speed. Given its still rapid velocity, however, it would continue to collect hydrogen via the ramscoop for several more months. As this source decreased, supplemental hydrogen from the reserve tanks would be used. The starship would continue to decelerate until it reached interplanetary speed and entered the neighborhood of Epsilon Eridani.

As the ship maneuvered through the star's system, its sensors would probe any planets, moons or asteroids it encountered, as well as the giant dust ring that had been known about since the early twenty-first Century. Planetary gravities would be assessed, surface features would be mapped, and atmospheric temperatures, pressures, and compositions would be analyzed. The main target, of course, was Protos, the fifth planet out from the star and situated comfortably in the habitable zone. If it indeed were judged to be suitable for colonization, the starship would orbit the planet and crewmembers would embark via the four shuttles that were docked in the wheel arms. At long last, humans would set foot on a planet orbiting a distant star.

32. Protest

Ethan hurried along the Main Corridor toward the Ballroom. As Chief of Security, he needed to be there should the protest get out of hand. As he came into the room, a moderately sized crowd stood before the stage. At the podium was Neil Jones, the leader of the Forever Space movement. He was attired in a blue jumpsuit that appeared baggy on his spare physique. With his short salt and pepper hair and preference for thick glasses over laser surgery, he conveyed a passive intellectual look that masked his rigid authoritarian personality. Beside him stood Ethan's old nemesis from his basketball days, Rufus O'Malley.

"We must do something now!" Jones declared. "The deceleration must stop, and we must give up the plans to colonize this star system. Who knows what we'll experience on Protos? Famine and flood? Disease-carrying microbes? *Protos 1* is our home, our life. We control the environment, we have enough food for our population, there is no disease to speak of. Our life is comfortable, and we can keep going to the end of the universe. Forever Space!" he chanted, "Forever Space! Forever Space!"

A few people in the crowd took up the chant. But one woman in a trim, fitted shiny jumpsuit moved forward.

"What about the Protos Mandate and the original charge to the Founder Generation? We are here to advance humanity by colonizing a distant exoplanet. We can't expand our population and resources by staying on *Protos 1*. We need to be on a planet. Expansion to other star systems will ultimately save the Earth. According to messages we have received from the Space Alliance, our periodic progress reports are helping them plan future missions ..."

"What do we care about the Earth?" Rufus interrupted. "It's only us out here. We should think of ourselves first."

"That's right," Jones added, "and remember that our species made a mess of things on Earth and polluted our Solar System with its expansion. We bring death and pollution wherever we go. We should stay in space, enclosed by our comfortable starship, leading our lives in peace and harmony."

An Elder stood up. "If we stay on the starship, we increase the danger of fostering the action of recessive genes. That won't be good for any of us."

"The gene issue is overstated," Jones replied. "With our genetic testing and modification programs, we can control for bad pairings and take care of any problems that may result."

"I don't mind saying that I'm afraid of terrestrial living," said a young man with a shaved head. "All that room, all that sky—I've seen holoimages of Earth before the Great Pollution, and it's pretty scary. The horizon goes on forever. There's no sense of enclosure, of being safe from the elements."

Several people murmured their agreement.

"But we'll get used to it," said the woman in the shiny jumpsuit. "Humans can adapt. We need some adversity to grow and develop. Life is too easy on board the ship. Science and technology are beginning to stagnate. There's no need to move forward, and that's dangerous for us."

"Dangerous?" retorted Jones. "We haven't developed any new science or technology because we don't need any. Life is good here. We have understood how to make this ship work for us, to create an environment that supports nearly 240 people in comfort."

"What about the Sleepers," asked an elderly man in the crowd?

"What about them?" said Rufus. "We don't have space and food to permanently house and feed another 40 people who have been trained for ground living and have no skills to use in running a starship. We either leave them asleep, or if they use too much energy to be kept in S-A, we turn off the juice and feed them to the Converter."

Howls of protest erupted.

"I have a relative in S-A," yelled an Elder. "I've thought all my life about meeting her, about learning from her what it was like in my family and in the Solar System before launch. We promised these people that we would activate them at Protos. We can't renege on our promise."

"Why not?" said Jones. "You never knew this relative. She's just a ghost from the past. We know all we need to know about the Solar System from the holocomputers, from the robotic teachers. Besides, that history is not relevant to us here in deep space on board a starship. We are the pioneers setting up a new human history where planetary existence is unnecessary. That's what's relevant today."

Pausing a moment for dramatic effect, he then continued.

"I've downloaded a petition on the starship master computer demanding that the Elder Council take up the issue of bypassing Epsilon Eridani and continuing on into deep space. It asks that all clauses of the Protos Mandate pertinent to colonizing Protos be stricken and that only those clauses remain that have to do with maintaining our society on board the starship. This petition will appear in your personal holocomputers, and all Citizens, Elders, and Grandelders must vote either yes or no by midnight tomorrow. Per the Mandate, if this petition receives 33 % or higher approval, it must be considered at the next Council meeting, which is next Tuesday. We of the Forever Space movement ask for your positive response."

He gathered his notes and left the podium, with Rufus close behind. Several Forever Space advocates in the audience followed, with the rest of the crowd dispersing in loud chatter.

Ethan remained behind until all but a few other security guards had left. He was glad that there was no physical altercation.

He was reminded of the Returnee Mutiny. *There's always some wacky extremist group wanting to change the Mandate*, he thought.

He then caught himself and smiled as he remembered the conviction that he and Sarah had felt, some 28 years before, about changing the Mandate's pairing rules

I suppose some would have said that Sarah and I were wacky extremists at that time as well, he thought.

But the Forever Space group had grown to include a sizable minority, enough to make their case known and to perhaps sway the Elder Council. Ethan had to agree that their comments made some sense. The needs of Earth and the Solar System seemed like a distant issue from the past that had little relevance to their current situation. But as the Security Chief, he was bound to defend the dictates of the Mandate. *Unless they're legally changed by the Council,* he thought as he exited the Ballroom.

33. Awakening

Captain Jennifer Bernstein listened intently to her Cryogenic Engineer as he explained the situation to her and the other officers who were assembled in the Control Room.

"Bill, are you telling us that several of them are already dead?"

William Monroe looked at his childhood friend, thinking to himself how the thirteen years she had spent dealing with the stresses of command had not diminished her beauty, despite the presence of wrinkles now beginning to form around her hazel eyes and the appearance of gray streaks in her naturally brown hair. He had always loved her, dating back to when they played together in the Alpha playground. When she married someone from the Beta District, he decided that his only desire for a mate had been taken and that no one else could replace her, so he received permission from the Council to remain a bachelor and devote his life to the S-A section. Unfortunately, he had bad news to convey today.

"Yes, Captain. When we instituted the revival procedure on the first ten Sleepers, three showed no evidence of brain function. We are able to keep their hearts beating artificially, but it is of little use. They are brain dead. We plan to take them off of life support."

"Dr. Zubkov, do you agree?" she asked.

"Yes, unfortunately, Captain. We are just using up energy. I see no hope for the three of them."

"What happened, doctor?" she asked. "How did they die?"

"We aren't sure. We've always known that the brain is the trickiest organ to freeze. Its metabolic needs are high, and some intravenous chemicals poorly penetrate the blood-brain barrier. This may have happened with some of our cryoprotectants at the time they were administered to the deceased during the original S-A procedure. In these three individuals, we extracted some brain tissue and analyzed it. It showed the presence of ice crystal formation. Although this could have happened during the revival procedure, we think that it likely occurred during initial cooling."

"Or the problem might have been related to the power outage we experienced in 2462," Bill interjected. "The deceased's cryopods were located in the area that was affected. Maybe this somehow negatively affected the supercooling process or the action of the cryoprotectants or even the nanobots—we simply don't know."

"So it seems that these three people have been dead for decades," the Captain said.

"Probably," Zubkov responded. "We couldn't measure brain function when they were in S-A, so we couldn't determine for sure what their status was. But it's a moot point—they could not be revived."

"Who are these three?" the Captain asked.

Bill consulted his wrist holocomputer.

"Jane Morgan, botanist; Andrew Alsufi, soil chemist; and Philip Jackson, geologist."

"How sad. I guess we'll have to check the roster and either cross-train some of the other Sleepers who have similar backgrounds or computer educate some of our Citizens to fulfill the tasks of the three people who died. Hopefully, we won't lose anybody else in the next group that you revive."

"We could be in trouble if we do," Zubkov said. "Not only would we lose skills necessary for establishing our colony on Protos, but we would lose some enhancement to our gene pool. We had counted on the sleepers to pair off with regular crewmembers after they woke up. This would add novel genetic alleles to counter the genetic drift and decay in heterozygosity that we have experienced in our closed population over the generations."

Bill looked forlorn. "I feel awful about this. I wish I could have done something for these three."

"Fortunately," interjected Zubkov, "the seven that awoke appear to be in good shape. One has some confusion that seems to be clearing as she becomes more active. Another had cardiac arrhythmias in the first hour after awakening, but those have gone away. Of course, nearly all have the characteristic retrograde amnesia for the few hours prior to the S-A induction, but this was expected. All in all, the survivors are doing well."

"That's good news, doctor. Well, I'll let you get back to work reviving the remaining Sleepers."

As the group began dispersing, the Captain pulled Bill aside.

"There really was nothing more that you could have done about this. You were not around at the original freezing and couldn't have prevented the power outage in 2462. Plus, long-duration total body S-A was relatively new when we launched. I think it's remarkable that so many Sleepers have awakened without problems after such a long time in a cryogenic state."

"I suppose so, Jennifer. Still, I'm responsible for cryopod maintenance, and it's hard to accept such a high death rate. These were our future colleagues and friends, people with special skills for a colony on Protos that we may not be able to replace. And the death rate could climb as we try to revive the remaining Sleepers."

She took his arm.

"Bill, death is hard to face. When my husband died a few years ago from the hydrogen flow accident, I felt responsible as Captain. Maybe I shouldn't have sent his team into the central core to check on the connections to the deceleration rockets. They likely were intact, but the protocol asked that we do this in preparation for the later firing. I could have overridden the protocol, but didn't. Luckily, my husband was the only one to die, but I still have nightmares about it."

"Protocol is protocol, Jennifer. Those connections needed to be kept open, or there might have been further damage when we converted from main to deceleration flow. It was not your fault. You did what you had to do."

"Nevertheless, it was tough telling my son that it was his mother's order that led to his father's death."

Bill took her hand in his.

"I understand, especially now, what you felt and are still feeling. For me, it's hard to stop thinking about the Sleepers who didn't survive. But I guess we need to learn that sometimes we can't control what happens in life."

The two of them locked eyes, then they separated when one of the officers called Jennifer over to review some navigation figures. As Bill walked away, he wondered if the intensity of their mutual gaze reflected more than just an old friendship. He hoped so.

34. Deliberations

Of 171 adults who voted, there were 65 yeas in favor of the Forever Space Petition, and 106 nays. At 38 % votes in the affirmative, the petition had enough support to be placed on the next Elder Council Meeting agenda just before the holidays. The meeting was held in the Ballroom since most crewmembers were likely to attend, and indeed nearly 200 people were in the room. Ethan came into the Ballroom shortly before the deliberations began and took a seat at the back where he could watch everyone.

The Councilors sat at their table facing the audience. Elder Council Chairman and former Captain Samuel Markman called the meeting to order. After dispensing with some routine matters, the pros and cons of the petition were considered.

Councilor Yvonne Brooks summarized the viewpoint of the petition's advocates.

"This petition reflects the opinion of many of our citizens who are happy with the status quo. They believe that they are true spacefarers whose home is the starship and whose world is the universe. Some are anxious about the life we may lead on a planet, with possible famine, unpleasant heat or cold, windstorms, illness from native microorganisms, even attack from any advanced life forms we may find. Our starship is a known and safe environment, it meets our needs, and there is no reason for us to leave."

Elder Ward Brown leaned forward toward his microphone. Several people expected the former pilot to support the petition, but he surprised them.

"The petition violates the Protos Mandate and our whole reason for being here. It's also a slap in the face to people on Earth and the Solar System who supported this mission over a century ago and saw it as a way to improve the condition of our species. I didn't devote my life as a pilot to take us to Nowhere. I had a mission to accomplish, and I want to see it completed before I fry in the converter."

"But," interjected another Councilor, "rules can be broken. We have amended the Mandate on several occasions to meet the current needs of the crewmembers. We can certainly do it again if that is the will of the Council."

Markman looked at him. "You know, things are not completely ideal on board this ship. People have to obey strict laws in order to keep the population at a sustainable level. If we remain on this ship, we will become more vulnerable to recessive genetic pressures, or to a fire or accident that will take the life of a large number of our people. Also, with ease comes stagnation. Why, we haven't seen any new scientific breakthroughs or technological advances since the self-cleaning robotoilet twelve years ago."

There were snickers from the audience. Markman smiled and continued.

"If we need to deal with adversity on a new planet, so be it. We will grow and learn as we adjust and as our population increases. People can pair off with whomever they choose whenever they want. Citizens will be able to have as many or as few children as they want, even in their 20s or 40s without needing to respond to a quota and birth licensing system. Freedoms will expand. Maybe that's what some of us are afraid of! It's easy to be told what to do, but hard to develop your own sense of picking what is best from a variety of choices."

The discussion continued in this vein, back and forth. Ethan observed Neil Jones and Rufus O'Malley over to the right sitting with their cronies and shifting back and forth in their seats. Several were holding bags, which he thought was strange. As a precaution, he stepped out of the room and called for increased security on his wrist compuphone. He reentered the room and shortly saw several of his people appear and take a position near the doors.

The Council concluded its deliberations. Chairman Markman stood up.

"Well, ladies and gentlemen of the Council, we have had our discussion, and now it's time to vote. Let's give ourselves a few moments to think about the issues involved."

There was silence as the council members reviewed their notes or sat in quiet reflection.

"Alright, all in favor of the petition raise your hand."

Four hands went up.

"All opposed …"

Four different hands.

"A tie. Well, I will break it with a negative vote. The petition is defeated. On to Protos!"

"This is a sham! We demand a new Council," Jones screamed as he and some of his cronies stood up and moved toward the council table with raised fists.

"Order in the room!" demanded Markman.

With an arm signal from Ethan, several of his security guards moved toward the protesting group. Seeing the motion, Jones and others in his group began pushing and shoving people. Rufus O'Malley reached into his bag and pulled out a knife.

The security forces reacted quickly, and several lifted restraint rifles and fired their charges. Because the protesters were close together, all of them became entangled in fibrous nets that retarded their movements and stuck them together. Only Rufus managed to cut himself free. He looked around and spotted Ethan. He ran toward him, and the two scuffled. In the process, he cut Ethan's left forearm, drawing some blood. Ethan grabbed Rufus' knife hand and flipped him to the ground. One of the security guards hit Rufus on the head with the butt of his rifle to knock him out. The rest of the protestors were subdued and led out of the room.

Ethan looked at the Councilors. All were well, although a few were shaken by the action.

"Thank you, Chief Johansson," said a relieved Markman. "Your forces were timely and well trained to subdue this gang."

"Thank you, sir. That's our job. We'll lock them up in the new Jail Section pending trial."

"Yes indeed, Chief. I think their actions against the Council will result in a guilty verdict. They can serve their sentences helping us to set up our colony when we arrive at Protos."

35. Family

After securing the prisoners, Ethan went to the hospital, where he was treated and sent home. Upon arriving at his quarters in the Alpha District, he entered into his small family apartment. Sarah had arrived home earlier from work at the Hydroponics Section and was standing by the foodbot. Their 14-year-old son Jason and 12-year-old daughter Andrea were away in the commons library doing their homework, accompanied by their nannybot. Sarah and Ethan had elected to sign a traditional marriage for life contract and to raise their two children themselves. After approval from the Elder Council, they were assigned family quarters. Like all married apartments, this included a small, efficient living/dining room, a tiny bathroom/shower cubicle, and a bedroom for themselves cut into a wall with a privacy screen. In addition, their apartment included a centrally partitioned double bedroom for their two children. The nannybot was sent to the central robot battery charging area at night or during times that the family preferred to be alone.

As Ethan walked in, Sarah saw the bandage on his left forearm.

"By the stars Ethan, what happened?"

"Just a scratch from a knife. The Forever Space gang didn't like the Council's verdict and created a ruckus. Rufus managed to cut himself free from a security net and attacked me. I guess I'm still not his best friend. The docs fixed me up—it only took six stitches from the robosurgeon."

She moved toward him and gave him a hug.

"Rufus really carries a grudge against you, doesn't he?"

"Apparently so. I would have thought that he'd learned his lesson three years ago after cooling his heels as the first inmate of my new jail. Stealing food being saved for the Sleepers was a major offence."

"He's bad news, Ethan, a real sociopath. Being part of the Forever Space group shows that."

Ethan laughed. "Well, you have him all figured out."

"Hello dad, hello mom," Jason said as he and his sister entered the apartment. "What's for dinner?"

"The choices today from the foodbot are algae fritters or beans from the hydroponic gardens, soybean cakes, chicken slices, eggs, mixed vegetables, and the usual variety of juices. Help yourselves."

As the children made their selections, Andrea turned toward her father and glanced at the bandage on his left forearm.

"Daddy, I saw the news report on the library's central holovideo about the Council vote and the disturbance in the Ballroom. The holoimage showed the commotion and those crazy people being netted, but nothing else. Did you get hurt?"

"Just a scratch, sweetie. Part of my job."

"Before she went to her quarters to be recharged, the nannybot told us a little about the first space people, that they don't want to go to Protos."

"They call themselves Forever Space, love," Sarah interrupted, joining her children at the small table with her meal selection. "And it's not so much that they don't want to go to Protos as they want to keep flying on this ship. Forever. Are you eating dear?

Ethan sat down. "I had a snack at the hospital, so I'm not hungry."

He turned toward his daughter.

"Some people are afraid of landing on a planetary surface. They prefer the comfort of our existence on this ship. But the whole reason for our existence out here is to colonize a planet on a star different from our own Sun, to provide space and food enough for you and Jason to pair with whomever you choose whenever you choose. Or at least after you finish school. All of us will have more freedom living on Protos."

"Yeah," said Jason. "I'd like to have some space to run and climb and explore some new areas. At holochess club today, Jimmy told me that on a planet you can look up and not see any walls. Is that true?"

Ethan laughed. "I guess that's right. When your mom and I were courting, great-grannie Mollie told us that she remembered looking out the window on Earth at the night sky and seeing a few stars when the pollution allowed it, and during the daytime, she could see a few clouds and the Sun's disk through the haze. She also said that in olden times, the sky was blue, and thousands of stars could be seen at night. Maybe on Protos, we can have that ancient experience."

"Wow, that would be great," Jason said.

"Resource-willing, we will land safely and have the kind of life humans were born to live," Sarah said.

"So that's why we have to go to Protos," Ethan added. "Now that the protesters are secured, nothing should stop us. You kids will grow up on a new planet, so you had better get ready. I hope you're learning about Protos at school."

"We are dad," said Jason. "The roboteachers are focusing a lot on planetology and settlement-building. And today I learned that the name "Protos" comes from some ancient Greek language meaning "first," since this will be the first exoplanet humans land on."

"Yeah," said Andrea, "and today I learned about how to survive living around things called rocks and mud, where you can slip and hurt yourself. Imagine—a surface that's not smooth!"

"There will be many more strange things on Protos," their mother said. "But for now, finish your meal and homework before bed."

36. Grief

All was black, then gray, then colored. Figures swirled around her like ghosts, speaking tinny, distant sounds. Something or someone was moving her arms up and down. Her heart was thumping madly. She tried to make sense out of the world. First, everything was herself. Then, the space around her was a room with people moving about. A tube had been inserted in her nose and down her throat, then painfully it was pulled out. She gagged, then breathed. Her right arm hurt where something stuck her that was connected to a machine. Then a thought: *I am alive!*

Audrey Moran struggled to make sense out of the world. She realized that she had been in suspended animation but couldn't remember how it happened. She must now be waking up. She felt cold, nauseous. She coughed and was able to weakly move her left arm. Someone was moving some sort of vibrating machine over the blanket that covered her wet body. She felt life-giving warmth coming back.

"Aaadreee," said a voice. "Aaudrecy … Audrey! You are fine. You are awake."

She dragged her eyes and head to the right. She was greeted by a foggy image: the brown eyes of a pleasant looking man with thinning light brown hair.

"I am Dr. Zubkov. You are in a cryopod on the starship *Protos 1*. You are waking up from suspended animation. It is the 14th of December, 2551. You have been asleep for 106 years. We are decelerating toward the planet Protos in the Epsilon Eridani system."

She vaguely heard what he said, as if she were in a tunnel hearing the echo of an oncoming car.

Awake … 106 years … Protos—yes, going to Protos.

Her eyes focused, and the man's image came in more sharply. She felt herself smile.

"Urghh. Aaaliiiive," she heard herself say.

"Yes, you are alive and well," he responded."

He turned and spoke to a woman next to him.

"Slow down the IV—she is responding well."

He directed a light into her eyes as she winced. He felt her wrist. He put a small instrument over her chest, and numbers appeared.

"Pulse regular at 68, EKG normal with only one premature ventricular beat, temperature still low at 36.2 degrees Celsius, but rising."

He turned toward her again.

"You are in star-shape," he said.

"What? Something is wrong with my body?"

He laughed.

"Sorry, that's just a slang term we've created around here while you've been in S-A … that is, in suspended animation. In Universal English, it means that you're shining brightly, that your physical vital signs are normal. You're doing great, Audrey. But don't push things. We'll have you up and about very soon."

Over the next hour, people massaged her and moved her arms and legs, fingers and toes, and, gently, her head. They told her over and over who she was, where she was, and the date. One man reviewed for her the progress of *Protos 1* over the past century and reminded her of her background and training and value to the mission as an astrobiologist. As she took this all in, she began to remember things.

Yes, I volunteered to go on this mission. My mother was sad but proud. She was in politics and helped this mission to launch. I was put to sleep in S-A. Me and Phil. Phil! Where is Phil?

"Where is Phil?" she asked hoarsely.

The man reacted in surprise. He turned and whispered something to a colleague, who went over to speak with Dr. Zubkov, who was attending to another person waking up in a nearby cryopod. He came over to her.

"Do you mean Philip Jackson, the geologist?" he asked.

"Yes, where is he?"

"I'm sorry, Audrey, but he didn't make it. He didn't wake up. He's dead."

Dead? she thought, *How could he be dead? I am alive. He is the only one I know well here. These are all strangers.*

She looked around as fear and tears came to her.

"How is he dead?" she whispered. "What happened?"

"We couldn't revive fourteen of the Sleepers. We don't know why, exactly, but they didn't live through the S-A procedure. But you and twenty-five of your colleagues made it. You will have friends on board from your generation. You will make new friends as well. We are all looking forward to having you Sleepers join us. Together, we will colonize a new world."

The tears kept coming.

Phil is dead. Oh my God, he's dead! Who are these people? How can I trust them? I'm all alone.

She started screaming in terror. She felt isolated, that she didn't belong anywhere. How could she survive?

"Nurse, add 5 milligrams of Calm-doz to her IV, stat!" Zubkov ordered to the woman at his side. He then turned back to Audrey.

"You will be fine. We will help you adjust to your new world. You will make new friends. Everything will be OK."

She looked at him. He seemed kind and concerned. She felt herself calming down. She gave him a tentative smile—he smiled back reassuringly

Maybe it will be all right, Audrey thought as she drifted off to sleep.

37. On-orbit

The New Year was greeted with great fanfare. Not only was this the final year of the transit to Protos, but all the new Sleepers were introduced and welcomed to the group. The expedition now numbered 67 Learners, 65 Citizens, 63 Elders, 45 Grandelders, and 26 former Sleepers. With the tipping of champagne and the movement of the master chronometer in the very crowded Ballroom to read '2552/01/01/0000', all class distinctions were declared ended by the Captain, and all 266 future colonists of Protos celebrated as one group.

Over the next few months, the starship slowed to interplanetary speed, and the deceleration engines were turned off. The ship continued to coast as it encountered the Epsilon Eridani system. The star's vast outer dust disk presented a collision threat, so the ship entered at an angle above the plane of the disk and the star's retinue of heavenly bodies. A number of planets were encountered and probed as the starship continued inward in the direction of the orange star. None of these planets proved suitable for colonization, since they were too cold and either had thick layers of poisonous gases in their atmosphere or no atmosphere at all. The first of two asteroid belts was encountered, which presented possible mining opportunities for the future. Then, 3½ astronomical units from the star, came the giant planet that had been known about since the year 2000. Jovian in appearance, it was over 1.5 times the size of Jupiter and surrounded by several large moons that were noted to be possible colony sites in the future after Protos was settled. Following another asteroid belt and two more planets, Epsilon Eridani's habitable zone was reached. At last, at a distance of 98.7 million kilometers, or .66 AU, from its star, Protos was spotted. It was calculated to revolve around Epsilon Eridani in 240 days.

Captain Bernstein ordered her navigator to use the yaw, pitch, and roll thrusters to plot a course that would put them in a stable orbit around their final destination. The starship maneuvered by the planet's two rocky moons, one called Alpha and the other Beta in honor of the former starship clans. It entered into orbit around Protos on July 1, 2552. This was followed by a great celebration on board the starship. Humans had at last crossed interstellar space and reached a distant planet revolving around a distant star. A radio message was sent back to Earth and the Solar System declaring their arrival for the historical records back home.

As the ship orbited Protos, everyone crowded around the holoscreens and windows to view their final destination. Epsilon Eridani bathed the planet in a pale orange light, but geographical features could be made out in the gaps between the moving clouds. Most of the surface appeared to be ocean. There were three large island continents. The one in the north was roughly circular in shape, with a thin indented area along its southeastern boundary. The middle continent straddled about a third of the equator and was shaped like the hull of a long boat, with the thin bow-like section pointing up to the northwest toward the indentation in its neighbor. The southern land mass was broadly triangular, its bluntly pointed top almost touching the southeastern boundary of the middle continent. A contest was held to name the three continents. The winning names were suggested by the starship's linguistic expert and were taken from their Greek language equivalents: Vorios for the continent in the north, Mesos for the middle one, and Notios for the southern land mass.

The ship's instruments and probes verified that the planet had Earth-like features. It was a rocky world, with an atmosphere that contained 82 % nitrogen, 13 % oxygen, 2½ % carbon dioxide, and the rest argon and other gases. Fortunately, its rotation rate was almost exactly 24 hours, in line with human circadian rhythms. With a radius of 6048 kilometers, it was smaller than Earth and had a bit less gravity, but it had a magnetic field that suggested an iron core. The relative humidity in the atmosphere was high, even down to ground level. Seventy-eight percent of the total surface was ocean, and additional water was located in the lakes and rivers.

Under telescopic observation, the centers of the continents had volcanoes and vast plains, with occasional craters that gave evidence for past meteoric bombardment. Various mountain ranges and undersea ridges were detected, and along with the lock and key appearance of southeastern Vorios and northwestern Mesos, geologists on board speculated that there had been tectonic activity. Around the peripheries of the continents near the oceans were rims of green, with curious brownish-yellow patches here and there that could be detected under very high magnification. The area of green was especially prominent in the bow of Mesos. There were no animals seen or signs of advanced intelligent life, such as artificially-made living structures or lights at night.

Over the next three days, preparations were made to the four shuttles to transport the future colonists to Protos. It was decided that for the first landing, two shuttles would be used for small landing parties to explore the surface. The tip of Mesos' bow was chosen for the landing, in part to explore the green areas, and in part to set up a desalination and water purification plant in its adjacent ocean. The two landings were scheduled for July 4th, humanity's new Independence Day!

38. First Landing

Captain Bernstein decided to command the first shuttle down to the surface, leaving her first officer in charge on the starship. Her shuttle launched at 0900 hours, followed fifteen minutes later by the second ship. Although capable of carrying three dozen colonists each, only a third of this number were on board the two shuttles in this first exploration of the planet. They both spiraled their way down toward Mesos. Spotting a relatively level area near the ocean, Bernstein directed her ship to land.

At touchdown, she announced: "After 107 years we have reached our destination. Bill, how safe is it for us to go out?"

Reviewing his instruments, Bill Monroe responded: "Well, Captain, it's a bit cool out there, about 8 degrees Celsius, and misty, with a relative humidity of around 85 %. And we may be a bit light-headed with the low oxygen and the relatively high carbon dioxide. But radiation levels are tolerable, and I don't think we'll need any breathing apparatus or protective gear other than our jackets if we take it easy."

"Excellent, but everyone, fasten a portable oxygen mask to your utility belts, just in case. Alright, let's get ready to go out."

They began collecting their gear as the second shuttle landed nearby. Thirty minutes later, the two shuttles opened their hatches, and the crews disembarked.

Reflecting the light from Epsilon Eridani, the sky and ground had a slightly orange hue. To the west, the nearby ocean glistened and moved with rolling waves. In the other directions, the land was flat, with mountainous outlines spotted far away in the mist. Green, dripping plant-like material covered much of the surface, with patchy brownish yellow areas several feet in diameter located here and there.

Bernstein turned toward her astrobiologist, the only one to have survived the Sleeper awakenings.

"What do you think of the surface material, Audrey? Do we have some plants here?"

"It would appear so," she said as she walked over toward the green area. She put on some gloves and carefully scooped up a moist sample, turning it over in her hands.

"It looks like some sort of algae that have learned to adapt to living on land. This is understandable, giving the high humidity and surface moisture."

Putting the sample into a sterile tube, Audrey then looked up.

"Those brownish-yellow areas are interesting. They also appear botanical in nature and seem to be interacting with the underlying green material. I wonder if they're feeding on it?"

She walked over to the nearest patch, pulled some of this material off of the green mat, and examined it with a magnifying glass.

"Could this be some sort of fungus?" she said, mainly to herself. She put a sample into a separate tube.

"Alright everyone," said the Captain. "Let's unpack our geological and meteorological equipment and set up camp. We have a lot to see and explore."

While the crewmembers went into action, Audrey placed the two samples in a shoulder bag and started walking toward the ocean. Five minutes later, she arrived at the shoreline. Except for the orange tint everywhere, it looked like the shorelines she had remembered from Earth before she went into S-A. After watching the waves roll in for a moment, she went over to a rocky area that contained tide pools. Examining them, she saw no fish or large multicellular animals, but there was some slimy green matter that looked like more algae. After collecting further samples, she tested the water and found it to be similar in its chemical makeup to the oceans on Earth.

Looking toward the horizon, she thought: *I wonder if there are any life forms swimming out there?*

She stared for a while, and even waded out a bit into the water, but she saw no animal-like organisms.

This planet is probably less than a billion years old, she thought. *It took much longer for advanced plant and animal life to evolve on Earth. Why should I expect to see anything advanced here? But Earth had a tumultuous beginning, with numerous impacts from small bodies in its vicinity. Maybe things are different here—it certainly seems peaceful.*

She was reminded of the life forms in her home Solar System: the silicon-based crystal creatures on Mercury, the primitive methanogens living in the hot spring vents in caves on Mars, the multicellular jellyfish-like organisms in the waters of Europa, the controversially alive hydrocarbons found deep in the cold ethane pools on Titan. Life took on many forms, although to date it appeared that only Earth had produced intelligent life. But what about the samples she collected? Would they be alive? And what was in the ocean before her and in the caves and crevices in the distant mountains behind her?

There's much for me study here, she thought, *like I had hoped when I left home. But I didn't think I would be the only astrobiologist alive! I must train some people in case something happens to me.*

She thought about her last conversation with her now-dead parents, how her mother cried at her departure, even though her mother was partly responsible for the expedition as chair of the space committee that funded it. Her father had been more stoical, but even he barely suppressed tears as he proudly hugged his daughter. She also thought of her brief lover, Phil, who had not survived the S-A procedure and would never get a chance to realize his dream of living and working on a new planet.

I guess I'm the lucky one, she thought, as she took one more look at the vast ocean, turned, and made her way back to the base.

39. Town Hall Meeting

Over the next few weeks, equipment, supplies, and all of the people except for a skeleton maintenance crew who remained on the starship were shuttled down to the surface of Protos to set up the colony. Most of them became fatigued easily in the low oxygen, so the work was slow, but with the help of the roboworkers it all got done. Everyone pitched in, even Neil, Rufus and the other Forever Space group who were drafted into the process as part of their parole. The prefabricated buildings went up quickly, and algae aquafarms were put into operation. Materials were brought down from the starship to build a fusion reactor for power. In the meantime, high efficiency batteries and solar cells met the energy needs of the colony, despite the drizzly and overcast conditions. Food production began to increase as the soil was found to support the grain crops whose seeds had been unfrozen from the starship. Most of the colonists kept busy and cheerfully reveled in their new freedom to pair off with whomever they chose, conceive as many babies as they wanted, and experience relatively few restrictions that limited their activities.

But not everyone was happy. Several people, including Neil Jones, became quite agitated and anxious on Protos.

"Its agoraphobia," Vadim Zubkov declared at the October town hall meeting.

"What's that?" Andrea Johansson asked her brother Jason. Since the children were sitting in the front row with Ethan and Sarah, her question was heard by Vadim.

"Agoraphobia is a fear of open spaces," the physician responded. "We never saw this on *Protos 1* because everything was enclosed by the walls of the ship. But here on a planet, we have wide open spaces on the surface and a sky that goes on forever above our heads. This is a difficult experience for people who have grown up on a starship. The fear of seeing a true horizon probably encouraged some of our people on the starship to join the Forever Space movement."

Audrey raised her hand and was recognized.

"What can be done about this?"

Vadim looked at her and thought to himself how lovely she was, with her beautiful brown eyes and auburn hair. She had the look of a young woman, despite actually being over 140 years old. Although they had been dating, he realized that he wanted to spend more time with her in the future.

"Right now, this disorder can be controlled with tranquilizers and supportive therapy," he responded. "Hopefully, people with this condition will learn to adapt to planetary living."

"Thank you, Dr. Zubkov" said Captain Bernstein, who was also the Chair of the new Planetary Council. "Now we will hear a final report from our astrobiologist, Audrey Moran, on the life forms found on Protos."

Audrey stood up and brushed by Vadim as she made her way to the podium. The touch gave her a thrill, and she had to confess to herself that she found him arousing.

"Hello everyone," she said. "I don't have a lot to add to last month's preliminary report, but I have finished all of my testing and will tell you my final conclusions. The slimy green stuff along the shoreline and in the tide pools is similar to our algae on Earth, as is the green material around us that has adapted to life on land due to the moist conditions in our local environment."

Some people in the audience groaned at this comment. Many colonists found the constant mist and fog, frequent rain, and weak sunlight filtering through the clouds to be very depressing.

Audrey continued.

"Like Earth algae, the Protos equivalents are autotrophs, getting energy through photosynthesis and producing oxygen in the process. They are multicellular, and each cell has a nucleus and plastids with chlorophyll, but no roots or stems. Life here has developed much faster than on Earth, perhaps because the big Jupiter-size planet nearby, which many of us are calling Zeus, gravitationally scooped up most of the debris in the early star system, so there wasn't the kind of disruption caused by frequent bombardments from space that our native Earth experienced in its youth. And the algae seem to be working on our behalf by using up the high carbon dioxide and converting it to more oxygen, so time is on our side."

"What about the brownish-yellow stuff?" asked a man from the back of the room.

"It's different. It consists of organisms like our fungi. They derive their energy not from photosynthesis but from consuming the algae. They are like parasites, needing carbon-based matter that is decomposing, either naturally or via chemicals they secrete that can kill and absorb the algae in their vicinity. The curious thing is that each mass that we see is actually a plasmodium."

This comment was greeted by murmurs from the audience.

"What I mean by this," she continued, "is that each is a clump of protoplasm that results from the fusion of separate amoeba-like cells that have come together. As a result, they lose their cell walls to form a shapeless glob."

"Globs—that's a good name for them," commented Ethan from the floor. "So without the algae, they wouldn't exist, correct?"

"As near as I can determine, yes. I have looked and looked and haven't found any other natural source of food for them. They have adapted to the presence of the algae that have made it to land. The algae get nothing from them, but fortunately they seem to be thriving on their own, so the globs are happy."

"How do they find the algae and move on top of them?" Ethan asked.

"They put out extensions of their protoplasm in the direction they want to go, then flow into it," she responded.

"Like primitive pseudopods," Vadim commented.

"Exactly. I've noticed some of this movement in my studies, and it's very slow. Once they find their food source, they are happy to stay around it until they have extracted all of its nutrients, then they flow over to an adjacent algae site to continue eating. This process is barely noticeable unless one measures the movement carefully over time."

Captain Bernstein stood up.

"OK, Dr. Moran, thank you for your comments. Any other business?"

A tall man with angular facial features stood up.

"Yes, Captain, I have something to say. The rule that we only get one day off work every two weeks violates my belief system and that of the five of us practicing our religion. God allowed for a day of rest each week, and we don't see why this couldn't be followed here. When the mission started, the Protos Mandate allowed for religious customs to be accepted unless they threatened the safety of the crew. There were times during emergencies when we all had to work every day to survive. But now that we have settled on Protos, things should be more flexible. Our beliefs need to be respected."

"Yes," a woman blurted out, "and why can't we stop work for periods of thoughtful prayer throughout the day, like some of our forefathers did on Earth?"

A buzz went through the audience. Bernstein stood up.

"I realize that different people among us have different belief systems and customs. But the fact of the matter is that we are still a very small and vulnerable colony. We must keep working to produce food and other material for our mutual safety. Everyone must pitch in equally. Although individual beliefs and customs are allowed where they don't interfere with the public good, we are basically a secular society grounded in reason and the law. Except for the general biweekly day off, we adults need to work every day, and our children need to attend school every day in order to master what they need to know to keep our knowledge base intact. This is the will of the majority and the unanimous mandate of your Council."

Most people applauded and nodded their heads.

"OK, is there anything else?"

When no one commented, she adjourned the meeting.

40. Relationships

That night, an hour after the town hall meeting ended, Captain Bernstein heard a knock on her door. It was Bill Monroe.

"Hi Jennifer. Mind if I come in?"

"Of course not. Have a seat. Cup of tea?"

"That would be great."

She went over to the foodbot and punched in her order. The tea came out almost immediately.

Receiving it from her as she sat down across from him, Bill said: "I thought you handled the meeting very well tonight, especially the issues raised by the religious groups."

She rubbed her eyes. "It seems like everyone has a special agenda. If it's not religious beliefs, it's some kind of cultural issue, or gender issue, or housing issue. Everyone is having sex at will but starting to balk at having a lot of babies, which we need in order to grow our population. I know we can start thawing the frozen embryos we brought with us, but the number is limited. People seem to forget that we're on an alien planet and trying to survive, and that the good of the colony is paramount. It seemed much easier on the ship."

"Maybe that's the problem."

"What do you mean?"

"We were all under the Protos Mandate on the trip out, with clear directions and rules, and a very controlled social environment. There was no true democracy. Now the restrictions have been lifted, and everyone is seeing what it's like to have more freedom. For example, the fact that everyone 16 years and older will be able to vote for the members of the Planetary Council in the next election. That's new. On the ship, a group of Elders picked the members of the Elder Council from within their Echelon, and up till now you have appointed people to serve on the Planetary Council. Things will be much different in the future."

She took a sip of tea. "Maybe you're right. These may be growing pains for our people. But it sure makes it hard to govern with so many special needs. Anyway, I won't be dealing with these issues anymore."

Looking surprised, he asked: "Why?"

"I've decided not to run for the Council again. I've spent most of my life training to be a leader, and too many years dealing with command as the Captain of *Protos 1*. I'm ready to give up the reins to other people."

"What will you do?"

"Probably devote more time to my hobbies: computer programming, three-dimensional holochess ... things like that. I may even take on non-hydroponic gardening to see if I can make something grow in real soil. In terms of service to the colony, I'd like to become a member of the surface rover exploration teams. We have a whole world to survey. Of course, if the decision is ever made to deorbit *Protos 1* and send it out to further explore Epsilon Eridanis' planetary system, I suppose that people would want me to be involved with this activity as well, so I'll have to keep sharp and review my captain's job description from time to time."

Bill laughed at her comment, but then became serious.

"I guess these activities will take you away a lot."

"That will be one part of the job, certainly."

He took a deep breath. "I had hoped that maybe you and I could do some things together, now that we're all settling in on this planet. You know that I like holochess, and the ocean here is spectacular. Most of us have never seen an ocean before, and I never seem to get tired of taking walks on its shore."

She smiled. "I'd like that. You are my oldest friend, and I would like to spend more time with you that's not involved with some crisis or engineering issue. Yes, it would be great to do this. Maybe we could have a "date"—go out to dinner to the Colony Restaurant or have a drink at the Protos Bar."

"I could even pay, like they did in the old days," he said.

They both laughed.

"I guess you could, except that we don't use dollar currency anymore. But you can spend your work credits on me," she said coyly. "You certainly have more accumulated than I do since you were appointed Chief Colony Engineer. You've really been working long hours since we've landed. In contrast, there hasn't been much work for me to do as the former Captain. The only work credits I've earned have been due to being Council Chair. So, I guess dinner's on you this time!"

"How about tomorrow night?" he asked.

"You've got a deal," she responded.

He took her hand, and she smiled.

****

The Protos Bar was busy that night after the town hall meeting. In the corner, Vadim and Audrey were sitting, nursing one of the bottles of Century Wine that had been brought along on the mission for special occasions. Vadim had won the bottle at the recent lottery drawing.

"This is much better than the home brew they're making here in the back room of the bar," Audrey said. "But what's the special occasion you spoke about?"

He leaned forward toward her.

"You know, we've been seeing a lot of each other the past few weeks, and it's been great. Tonight, as you were speaking at the council meeting, I realized how much I care about you and would like to spend more time with you. Maybe we could consider a more formal pairing, like living together. We could sign up for "new couple" status and ask for larger quarters in one of the new buildings near the Auditorium."

She smiled. "You mean you'd like to live with a much older woman? After all, I'm more than 100 years your senior!"

He laughed. "That crossed my mind. But you've adapted quickly to our modern culture. I even heard you saying "by the stars" the other day! My point is that I love you and want to be with you."

Her eyes watered over. "I've begun to feel the same way about you. I don't know where this is going, but I think we should talk about something. I'm kind of an old fashioned girl, and if we ever decide to have a baby together, there may be a problem. None of us former Sleeper women have gotten pregnant yet, and no one knows for sure if the S-A process has done any damage in terms of pregnancy or fetal development. You've raised some questions yourself."

"That's right, but they've been more hypothetical and based on the slow recovery of hormonal balance in both men and women after the awakening. But menses finally occurred in all the women, and by six months or so after coming out of S-A, hormonal imbalances generally have corrected themselves in most Sleepers. Plus, the sperm and egg tests we performed were normal. There have been no pregnancies yet, so no live births have come from Sleeper women. But you all haven't been awake very long, and not many women in your group have even tried to conceive, at least according to what my Sleeper patients have told me. But you and I know that your libido is certainly working!"

They both laughed. Then she became pensive.

"But Vadim, there still might be a risk. No one expected so many people to die in the recovery process, yet fourteen did. One was a friend of mine. I just worry about not ever being able to have a baby, or having an abnormal one."

"The frozen animal embryos that we brought all grew normal animals, so things will likely work out for the frozen human embryos and fertilized cells we brought along. Implantation is always an option. At any rate, as far as I'm concerned, I'm willing to take a chance and see what happens in our relationship."

"Me too," she said.

She leaned over and gave him a kiss. They finished their drinks and left the bar. They walked hand in hand down the main street of the colony toward her current quarters near the eastern algae growing vats. The street had just been paved over, so it was not as muddy as the side streets that were exposed to the dewy night. They entered her small quarters on the main floor of her apartment building, cooked themselves an algae omelet for a snack, drank a couple of local beers, snuggled to the music from her holoA-V machine, and then successfully satisfied their libidos as the night wore on.

41. Globs

On November 1st, 120 days after landing on Protos, the Planetary Council held a special meeting to address the feasibility of continuing to use Solar System Universal time on the new planet. The Councilors decided that it made more sense to match the time and calendar to local planetary conditions. Since the period of daily rotation of Protos was just a few seconds short of 24 SSU hours, there was no need to change the number of hours in the day. The Alpha Moon revolved around Protos in 20 days and the Beta Moon in 31, so the former time was picked as the duration of the new month for ease of calculation. This meant that there would be 12 Alpha months in the 240-day year rather than the awkward 7.74 Beta months. The Alpha months were given sequential letters from A to L each year, and they were divided into four weeks of five days each. The day names continued in the old manner (with Saturday and Sunday being permanently dropped, to the chagrin of many religious colonists). The changes were implemented a few days later when the planet was thought to be entering the 60 days of local summer. The date was arbitrarily picked to occur mid-year on Monday the 1st of G month. This day was subsequently celebrated each year as Protos Calendar Day.

****

Over the next week, the weather began to change. The clouds thinned, the mist and rain decreased, and the temperature began to top 15 degrees Celsius. The relatively warm weather coincided with a scheduled expansion of the colony's building space. This necessitated cutting back the local land algae, which was relatively easy since some of the plants were drying up due to the weather change. But this was not the case for the settlement plants in the eastern and western algae food vats, which were well watered and thrived as they received more sunlight.

As Bill Monroe was walking back to the settlement from the ocean after his routine inspection of the new desalination plant, he wiped the sweat off his brow and happened to glance over toward the distant mountains to the east, which were easy to see now that the mist had dissipated. He noticed that some of the surrounding fungal globs seemed fewer in number but larger than before.

That's odd, he thought.

Stopping to gaze at them, he noticed that some were extending out pseudopods in the direction of the settlement. They seemed to be moving toward the eastern algae vats on the periphery of the settlement.

He ran over to the Apartment District and knocked on the door to Audrey Moran's quarters in the new couple section. She answered with a smile, still thinking of a particularly nice breakfast conversation she and Vadim had had that morning before he left for the hospital.

"Hi Bill," she said. "What brings you my way?"

"Something is going on with the globs today. Some have merged together and are extending out pseudopods in our direction."

"What, you saw them move? That's unusual."

She grabbed her sample bag, and the two of them went over to the eastern part of the colony. Sure enough, the globs were very active, visibly flowing toward them and merging together when two or more made contact. In addition, some of them were growing vertical stocks, at the top of which bulbous areas were forming.

She took some pictures with her wrist compuphone and used it to send images to Captain Bernstein, Ethan, and some of her fellow scientists, calling on them to join her by issuing a Code 2 Alert.

"They certainly seem to be on the move," she said, looking at the globs again, "and changing their shape."

There was something about this activity that seemed familiar, something she could not quite recall.

Captain Bernstein and Ethan came running up, along with a couple of her colleagues. She began to brief them, and as she spoke she flashed on a memory of herself as a young girl in Oregon walking through a misty forest of Douglas Fir trees.

"Slime molds, that's what they're like," she said.

"What?" asked the Captain.

"On Earth, slime molds change their character when they are stressed or undernourished. They react by forming sporangia, stalks of spores that are released to create new mold cells and thus preserve their species. This sort of thing seems to be happening here."

"But why now?" asked Ethan.

"The change in weather has dried up their land algae food supply, and we've cut away some of the rest in order to expand our settlement. The relative heat is probably stressing them out as well—they like a cool wet climate."

"Why are they heading in our direction?"

"Because we are the closest source of food. We have algae in our hydroponic vats. Slime molds receive chemical triggers from their food sources, and they instinctively move toward them. Some scientists used to think they had a primitive type of intelligence. I think something like this is stimulating the globs to merge and move toward our algae."

As they watched, one of the sporangia on a nearby stalk split open at the top, and thousands of small particles were released. Some were shot out; others were picked up by the wind and dispersed. A few landed on a man tending some crops nearby. He paused, scratched himself, then started screaming. He ran toward the hospital, and as he passed nearby they could see blisters forming on his exposed skin. He ran through the Emergency Room door.

"By the stars, the spores are toxic," Ethan said. "We've got to get people away from the periphery."

As he commented, other stalk tips exploded out their spores, and the air took on a cloudy character where they were ejected. Other globs moved at an increasingly rapid flow toward the algae vats.

Ethan spoke into his wrist compuphone. Almost immediately, an emergency message was announced over the colony loudspeaker for people to evacuate the eastern periphery. He then called some of his men to bring welding flame torches to his location, and he contacted Sarah and told her to gather her staff and leave the hydroponic sections. While he was making these calls, people came running over toward him.

"Get inside," Audrey yelled. "if you can't get home, head for the Auditorium. Don't let any of the spores coming from the globs get on your body."

Another man screamed in the distance, scratching and rubbing his blistering skin. A nurse came running out of the hospital to tend to him, and she in turn was contacted by some spores, causing her to scream as she retreated back into the Emergency Room.

Ethan turned toward Audrey, Bill, and Captain Bernstein.

"You had all best get into the Auditorium yourselves. Things might get nasty out here."

As they departed, several security guards came running up carrying flame torches, and Ethan directed them to fire on the globs. One man headed toward the nearest organism, directed his flame toward its closest pseudopod, and it immediately flared up in flames and smoke. The glob seemed to want to withdraw from the fire and heat, as a new pseudopod formed on its side opposite the flame. Other close globs also began to move away. But globs farther out continued to release spores from their sporangia. Two of the guards began scratching and screaming, and Ethan decided that there was too much risk involved in continuing the attack. He called for a general retreat, and he and his guards all headed back to the Auditorium, along with other colonists in the area.

Safe inside, Ethan joined Audrey, Bill, and the Captain. They looked out through one of the Auditorium windows. Several of the globs had merged to become an organism the size of a transport truck. It moved over to one of the algae vats, flowed over it, and dropped on the moist plants within. It ceased moving, stopped shooting out spores, and seemed to settle down while it feasted. Other large globs were doing the same thing, and in half an hour, everything became calm again.

"Well," Ethan said, "it would appear that we've bought some time."

"Yes," Bill said, "But I fear that we'll lose a lot of our food supply. For now, the algae vats on the western side of the colony toward the ocean are safe, probably because the globs on that side have plenty of native algae to feast on that are in the wet soil near the water."

Turning to Audrey, Ethan asked: "So what happens when the eastern globs finish eating and get hungry again?"

"I suppose they'll produce more spores and want to flow their way through the settlement in the direction of our western algae vats. That won't be good for us. If they succeed, we could lose most of our food supply. We just don't have enough soil-grown plants or animals to support our colony."

"How much time do you think we have?"

"It's hard to say, Ethan. Right now the globs are quiet, with plenty of food available, and the water in the vats is keeping them hydrated. Maybe we have a day or two, or perhaps even more time. Our algae supply is not native to this planet, and the globs may reject some of it or metabolize it faster than their native algae. I just don't know."

"Now that they're quiet, maybe we can just blast them with fire or explosives," interjected the Captain.

"There are too many of them out there to deal with in this way," Audrey said. "And we don't want to excite them further to shoot out more spores."

"Plus, we'll destroy the algae and much of the settlement in the process," added Ethan. He thought a moment, then said: "You know, maybe we can make it unpleasant for them to bother us. Several of the globs avoided or moved away from the fire before we had to evacuate that plan. They seem to feel pain."

He turned toward Bill.

"Do we have enough wire to build a fence around the settlement, including the ocean side algae vats?"

"Yes, but how would that stop the globs? They would just flow through the fence."

"Not if the fence is electrified. One touch, and they would feel pain, like they did with the fire. Maybe that would cause them to avoid the fence and go around us as they continued toward the native algae near the ocean."

"Wow," said Bill, "that might work!"

"How fast could you build such a fence and link it up to our power generators?"

"With the help of the roboworkers, probably in two days or so."

"What do you think, Captain?"

Bernstein looked at Ethan.

"It seems like a good plan to me. I can't think of anything better. If it works, then we can protect not only ourselves but all the food sources located inside the perimeter."

"And the desalination plant and pipes bringing us fresh water would not be affected, correct?" Ethan asked Audrey.

"Correct," she responded. "The water is encased in pipes, and the globs would not consider them or the water purification equipment to be food."

"So," said Bill, "if we can keep them away from the settlement, we'll have enough food and water to survive long enough to come up with a more permanent solution, or until the weather changes back to its wet state and the native algae start growing again."

"OK," said Ethan. "Let's get started with this plan and hope that the globs will leave us alone for a while."

42. Barrier

Working around the clock, the colonists managed to put up the perimeter fence in 42 hours, fortunately before the globs had decided to move on and look for more food. Bill had it hooked up to some electrical generators, and the system was tested and deemed ready to go. Captain Bernstein called an emergency meeting of the Council that evening in their chamber in the Auditorium.

"Through our good work, we got the fence built in time. Now we wait and see how effective it will be in keeping the globs out. Ethan, what are the plans?"

"I have my security forces concentrated on the three sides facing the globs, especially the eastern sector. The side toward the ocean is also covered, but with fewer guards. Also, many of my people will be wearing old space suits that we've modified to protect them from the spores. They will have flame torches and should be able to get close to the globs without being harmed."

"Excellent. Any new information about those spores?"

Vadim rose.

"They produce a kind of allergic reaction in non-native organisms. Some of the sheep and cattle we raised from the frozen animal embryos from Earth were hit by spores, and the animals had the same reaction as our people:

itching, then a hyper-allergic local blistering, then a more systemic reaction leading to cardiac arrhythmias. Most people we got to early enough recovered with massive steroids, but two died from the cardiovascular collapse."

"Audrey—anything from you?"

"The spores seem to be diploid in nature, which means that they have all the genetic material they need to transform into glob cells after they land on a source of food, the favorite being the green land algae. But a few spores landed on some of the soil crops we're trying to grow. Their spiky, sticky surfaces easily attached to them, and they began to grow. So they show some ability to diversify onto alien organic material."

"Well that certainly is not good news," Bernstein said.

Audrey paused a moment to consult her notes, then continued.

"I think the problem we have with the spores is incidental to the survival needs of the globs. They have no natural enemy and so have no need for defense. The spores are purely a reaction to times of stress, where the globs try to preserve their kind by setting off a massive spore production."

"Lucky us," Bernstein said sarcastically. Turning to Sarah, she then asked: "What about our food supply? Is it protected?"

"Yes Captain. My staff and I have surveyed the area. The fencing completely surrounds the crops and algae vats, and it appears intact. In addition, where possible we have erected tents and other coverings so that the spores cannot land on our plants and farm animals."

"Good work, Sarah. OK, let's move on to other business. What's new with the fusion reactor?"

As one of the engineers stood up, Ethan perceived a vibration on his wrist compuphone. He stepped outside the room.

"Johansson here," he said.

"Sir, this is Sergeant Chin. We have a report of someone creeping around the eastern perimeter near the number 4 algae vat, one of the central ones that hadn't been reached by the globs. Officer Petrolli went out to investigate, and the intruder told him to back off or he would blow up some vats. I thought you should know."

"Thank you. I'll be right there."

Ethan hurried out of the Auditorium toward the vat. Three of his guards were squatting down by a tractor. Ethan joined them.

"Report please, Sergeant Chin."

"The man says he has explosives and plans to destroy the fence and the adjacent vats. It's Rufus O'Malley."

"Damn," Ethan responded.

He stood up and looked toward the perimeter. In the shadow by vat number 4 he saw a figure.

"Rufus, what are you up to now?" he called out.

"Is that you Ethan? You're still hassling me? I've had it with this colony, with all of you. If you don't give me access to a shuttle to get out of this hell-hole of a planet, I'll blow up the fence and as many of the vats as I can."

"Rufus, you know I can't do this. The shuttles are all under the control of Captain Bernstein. We need them to transport people and supplies to and from *Protos 1*. Besides, you can't fly a shuttle by yourself."

"I'm sure there will be other people who will want to join me. A couple of the Forever Space folks, for example. But don't give me any crap about this. I have plenty of explosives from storage, and I will use them."

Ethan thought for a moment and decided that further discussion with Rufus would be useless.

"OK, I'll contact the Captain."

"You have 30 minutes, then I expect an answer. And a hostage. How about Sarah? She and I should have hooked up anyway a long time ago."

Damn, he never gives up, Ethan thought.

He turned and headed back to the Auditorium. On the way, he called Bernstein on his wrist phone and briefed her on the encounter with Rufus. Upon entering the chamber, he saw her in deep discussion with the other councilors. She looked up at him.

"Ethan, we were just considering this new development. We all agree that there is no way we can give up a shuttle to Rufus. We've got to stop him. Any ideas?"

"I can try to get to him," Ethan said. "But if he in fact has explosives, it could be very risky."

"Do what you can," she said. "We'll support whatever you do."

Ethan left and went to the Security Building. He quickly put on one of the modified space suits and grabbed a flame torch. He then made his way to the vat area.

"No go, Rufus. The Council will not give you a shuttle. You had best give yourself up for the best of the colony and for yourself as well. If you blow up a vat or destroy our fence, and the globs come in, we all will be threatened."

"You and the Council made your decision," was the response, "and now you must live with it."

Rufus came out of the shadow and ran toward vat number 4. As Ethan and his men sprinted toward him, Rufus placed a packet down and seemed to activate a switch. He then ran away.

A few seconds later, there was an explosion. The vat shattered, pouring water and green slime to the ground. The hydroponic tubes supplying additional water and nutrients broke, spewing fluid into the air. Luckily, the explosion did not quite reach the perimeter fence, which remained intact.

Ethan ran after Rufus, when a laser beam sizzled the insignia on his shoulder.

A laser pistol. He's armed!

Ethan dropped to the ground. He held up the flame torch and tried to point it at Rufus.

A dark cloud suddenly blocked the light from Alpha Moon. Ethan looked out toward the globs. One of them had been aroused by the algae that had spilled onto the ground and had begun moving toward the shattered vat. Furthermore, it had shot out a bolus of spores, some of which were attaching to Ethan's suit.

As he directed his men to pull back, he heard a scream. It was Rufus, who had no suit on and was in the direct line of the rapidly moving spore cloud. Hundred of spores reached him, as he scratched and began to run around crazily as if he were being attacked by a swarm of bees. Shortly he fell to the ground, blisters forming all over his exposed skin, then he stopped screaming and moving as he grabbed at his chest. Almost immediately, it was over, as he succumbed to the massive spore attack.

Ethan turned toward the oncoming glob. Spores continued to land on his suit, and two large pseudopods flowed toward him. He pointed his flame torch and fired over the fence. The creature immediately went up in flames. Other globs in the area that had been aroused by the uncontained algae came toward him. One caught fire from its flaming brethren. Another brushed against the fence. A spark went up, and it immediately retracted its advancing pseudopod. A third moved into the fence more directly, and the strong electrical charge immediately set it on fire.

Then a strange thing happened. The remaining globs stopped and retreated. It was as if they understood what was happening, that the fencing was dangerous and needed to be avoided. They stopped shooting out spores and went back to grazing on the algae supplies beyond the fence that they had captured a few days earlier.

"It's working," said Sergeant Chin, running up to Ethan.

"Maybe for now. Let's hope the fence holds and the globs leave us alone for good. Let's go get what's left of Rufus."

They moved in the direction of Rufus' body. It was a mass of blisters and blood where he had futilely scratched himself before he died. His face was frozen in a grimace of agony. Ethan felt sorry for him, even though Rufus had caused him trouble ever since their school years. He had been a person who could not abide by the rules of a society, and now he and the colony would no longer be a problem for each other.

Two guards joined him, one of whom placed Rufus' remains in a body bag, and they took him away.

43. Peace

Things were calm the next day. The globs seemed content to feast on the already captured algae without moving toward the colony. Audrey reported at a Council meeting that night that she thought the algae spill resulting from Rufus' bomb explosion had sent out some sort of chemical stimulation that triggered the globs to move forward toward a possible new food supply. At the same time, the explosion caused a stress response that stimulated spore release. This was compounded by the fire and the electrical charges from the fence.

"But why did they stop?" Ethan asked.

"I don't know," she responded. "Maybe the death and injury of some of the globs released some sort of chemical signal to the others that they should avoid the area."

"Do you think that they were consciously aware of the danger?"

"It's hard to believe that could happen. As near as I can tell, the globs don't have a complicated nervous system. But maybe there is some sort of stimulus-response intelligence going on with them. It could be instinctual. I'll have to study them further."

"Yes, indeed," Vadim added. "Despite the similarities of the algae to our algae and the globs to slime molds, we have to realize that they are organisms evolving on a planet around another star, and that their biosphere is quite different from ours. Maybe as we get to know them better, some of the mystery will be resolved."

The behavior of the globs two days later added to this puzzle. Almost on signal, they all started to extend out pseudopods and began to move. However, rather than come toward the colony, they proceeded around the perimeter of the fence until they arrived to the west, then they moved straight toward the native land algae that grew in the area between the settlement and the ocean. A new food source was being sought, but the colony was being left alone. Furthermore, no new spore stalks formed, and those that had already been erected were quiescent. The globs seemed hungry, but they did behave in a way that indicated alarm or stress as they slowly flowed onward toward new feeding grounds.

Audrey and some of her biotechs followed and observed them during their migration. They seemed satisfied feasting on the coastal algae. The weather had cooled some, and it was raining, which likely contributed to their contentment. Most of the spore stalks had regressed, and the large globs were beginning to separate into smaller pieces, much like they were before the warm period.

I guess we'll have some peace now, Audrey thought after an early evening glob observation period. She walked over to a cliff that faced the ocean. The waves were a bit more active due to a conjunction of the two moons, and a wind was coming in toward the shore. She never tired of looking at the ocean by day or in the light of the two moons at night. The surf comfortably reminded her of her native Earth. But she had a new home now, and a new life with Vadim. *And interesting life forms to study*, she thought. Contentedly, she turned and went back to the colony.

44. Celebration

The results of the year-end election were in. A new Planetry Council was formed, and Ethan automatically became its leader since he received the most votes in the general election. The swearing-in was to be held outside in the square on the first day of the Protos New Year. The winter weather was cool and drizzly, but there was no rain.

The first to be sworn in was Grandma Mollie, daughter of the first starship captain and the chief physician, who had recently turned 113 years old. She was the senior-most person in the colony thanks to the abolition of the mandatory death rule four years previously. Sarah sat in the front row holding hands with Jason and Andrea, who sat beside her. They were beaming at seeing their great-grandmother on the stage. She walked with a slight limp, refusing human support but accepting the aid of an old-fashioned cane that Ethan had made for her. She came up to the podium, where she was sworn in, then she turned toward the microphone.

"Well, you made an old gal very happy," she said. "I remember my mother and father wondering what it would be like to finally reach Protos. I'm still around and know what they never could know, that we are here, surviving, and sharing the planet with other creatures. They would be happy to see that their efforts were not in vain. I thank you for electing me, and I will do what I can to serve you and to make this a better place for all of us."

To spirited applause, she hobbled back to her seat. Then, in turn, the other Council members were sworn in. The last was Ethan.

"I think I speak for all of us Council members in thanking you for your trust. As the new Chairman, I will do my best to serve you and to steer a good course for us all. Humans have made it to an exoplanet at last, and I expect that there will be more to follow. Maybe they will come here, maybe somewhere else, but members of our species have become interstellar travelers."

"We have plenty to do here. The globs are quiet now, but we need to understand them better. We can't live behind an electric perimeter fence forever if we are to grow and prosper. We need to develop industries, expand our farming, and ensure the stability of our food and water supply."

He paused, looking up from his notes.

"Maybe we can figure out a way to coexist with the globs, perhaps even communicate with them in some way. After all, they are a native life form on this planet, and they seem to have an ability to communicate with each other. Furthermore, we must explore and map the planet to look for other areas suitable for colonization and to see if there are other native life forms that we haven't yet discovered."

"We also must explore this star system. *Protos 1* is in orbit, ready to go. As was the case in our native Solar System, the other planets, moons and asteroids of Epsilon Eridani might be tapped for resources that we lack here on Protos. Perhaps life exists elsewhere as well. We know that Zeus has many orbiting moons, and this giant planet needs to be explored more fully, as do the smaller planets in our new star system."

"But we need to work together, as one people. Individual differences and needs will be honored, of course, but our common goals must be supported by all. In the spirit of continuing the human species, our colony must not only continue, but progress and expand. This will be my goal as your new Council leader."

The crowd erupted in wild applause. The recent encounter with the globs had brought everyone together, and Ethan's speech seemed to summarize their determination to survive. Humans had indeed become citizens of the stars.

Part II

The Science behind the Fiction

A star map of the northern celestial hemisphere, which plots the location of the naked-eye stars in the northern sky along with images of their corresponding constellations, from Johann Zahn's *Specula Physico-Mathematico-Historica...*, published in 1696. Today, a plethora of planets have been discovered that orbit distant stars, but whether the heavens are also teaming with life remains to be seen. Courtesy of the Nick and Carolynn Kanas Collection; and *Star Maps: History, Artistry, and Cartography, 2nd Edition*, Nick Kanas, Springer/Praxis, 2012.

Challenges of Manned Interstellar Travel

Traveling to a distant star presents a number of challenges. First and foremost is the immense distance involved. For example, the nearest stars to us are in the Alpha Centauri system. The closest of these, Proxima Centauri, is 4.22 light-years away, which translates into nearly 40 trillion km (or 24 trillion miles). This is around 271,000 times the distance between the Earth and the Sun. Other stars are much farther away. These tremendous distances raise a number of issues related to methods of getting there, the long-term effects of time and space on the physiology and psychology of space travelers, and the chances of finding planets with life around a selected star. But before discussing the scientific and technological issues involved, let's take a look at how interstellar travel has been depicted in science fiction.

1 Science Fiction and Interstellar Travel

For the ancient Greeks, the stars were seen as being located in a finite sphere that revolved around the Earth, which was perceived to be at the center of the cosmos. This notion continued into the sixteenth century, when Copernicus moved the Sun into the center with the publication of *De Revolutionibus* in 1543. In 1576 the English scientist and politician Thomas Digges, a Copernican, published a diagram in a well-known almanac started by his father that depicted the stars as extending outward into infinite space. About the same time, the peripatetic Italian polymath Giordano Bruno was advocating the notion of a plurality of worlds, which stated that our Sun was but one of an infinite number of stars in a cosmos that supported an infinite number of inhabited worlds, some by intelligent beings. Bruno was branded a heretic and was burned on a stake for his ideas in 1600. However, the notion of the stars being solar systems with inhabited planets took hold, and it was discussed and illustrated in books and celestial atlases from the seventeenth century to the present day [64, 65].

The existence of life on heavenly bodies became the subject matter of science fiction writers as well. For example, in arguably the first science fiction story, Kepler's *Somnium*, published in 1634, the protagonist was transported to the Moon, where he encountered life forms engaged in a variety of activities [101]. However, Kepler was not an advocate of an unbounded universe and the existence of life throughout the cosmos. Indeed, most science fiction authors who subsequently wrote about space travel focused on our Solar System, particularly the Moon, Mars, and sometimes Venus (see for example stories by Jules Verne, H.G. Wells, and Edgar Rice Burroughs). However, in 1928 E. E. "Doc" Smith initiated the first of his space operas entitled *The Skylark of Space* [109]. First appearing in the pulp magazine *Amazing Stories*, it described the interstellar adventures of its protagonist whizzing through the cosmos in his faster-than-light space ship. The popularity of this work led to a number of sequels that are still in print today. In Smith's universe, most stars had planets, and many of these were inhabited by humans.

Both Charles Sheffield [107] and Simone Caroti [21] have reviewed the history of starships and science fiction during the twentieth century. Previous to this period, Caroti states that some 46 plays, short stories, and novels of spacefaring science fiction were written between 1865 (when Jules Verne's *From the Earth to the Moon* was published) and the end of the century. Between 1900 and 1926, another 74 works appeared [21, p. 21]. This period included articles involving ships traveling to distant stars by the "fathers" of the starship genre: Robert Goddard, the American rocket expert who in 1918 wrote a short manuscript describing giant space "arks" that transported people cryogenically cooled in a deep sleep to distant stars; Konstantin Tsiolkovsky, the Russian space flight pioneer who in 1928 wrote a paper describing the construction of enormous multigenerational "Noah's Ark" vessels to transport people to the stars over periods of centuries or millennia; and John Desmond Bernal, the British scientist who in 1929 popularized the multigenerational starship idea in his book *The World, the Flesh, and the Devil*.

Caroti outlines several eras in starship science fiction writing. The first is the Gernsback era (1926–1940), named for Hugo Gernsback, the editor of the first English-language science fiction pulp magazine, *Amazing Stories*, and later *Wonder Stories*. In the latter magazine, a multigenerational starship narrative appeared in September 1934, Laurence Manning's "The Living Galaxy," where humans engaged in multi-million year starship missions thanks to an anti-aging procedure that allowed them to live forever. Population control was established by periodically dropping people off to colonize planets along the way [21]. The story described how a group of people converted an uninhabited asteroid into a multigenerational starship and undertook a 4-million-year mission to confront a monstrous assemblage of galaxies measuring hundreds of thousands of light-years in size that were devouring matter in the cosmos. The staggering temporal and spatial dimensions in this story superbly illustrated the presence of the science-fictional "sublime" described by Csicsery-Ronay as one of the seven novums of science fiction as a genre [30], a characteristic found in many subsequent multigenerational starship narratives. However, relatively little was said in Manning's story about the psychological and sociological problems experienced by a group of people who interacted with each other through many generations over long periods of time as they traveled into the lonely outer reaches of space. Instead, the crewmembers were pictured as vigorous, inventive, and happy to be there, with no major personality or interpersonal difficulties.

A much more toned-down picture was painted in Don Wilcox's "The Voyage that Lasted 600 Years," which appeared in *Amazing Stories* in October 1940. Here, the crewmembers had a normal life span, and a succession of generations occurred during the ship's mission, with offspring numbers carefully controlled to assure a stable population. A person called the Keeper of Traditions was put into hibernation to slow down his metabolism and was awakened every 100 years to educate succeeding generations about the purpose and goals established by the original mission planners. However, his infrequent contact was ineffective, and over time the crew culture devolved into a condition of social decay and loss of technological know-how, with disease and political instability plaguing the population. The story described the tension between these psychosocial stressors and the Keeper's attempt to correct things according to the original purpose of the mission [21].

In looking at this and other stories of the Gernsback era, the stories reflected traditional American values of the time, with good guys and bad guys and relatively formulaic plot lines. They were a relatively inexpensive way to escape the stresses and strains of the Depression. Science fiction was seen as a guide for a better tomorrow, promulgating values in accordance with the tastes and expectations of the largely young male readership.

This began to change during the Campbell era (1937–1949), when John W. Campbell, Jr., took over the editorship of *Astounding Stories* magazine and renamed it *Astounding Science Fiction*. Perhaps reflecting the turbulent international scene leading up to and including World War II, science fiction stories became more attuned to real human concerns. Attention also began to be played to scientific plausibility and better storytelling. These trends were found in two related stories by Robert A. Heinlein, "Universe" and "Common Sense," which were originally published in *Astounding Science Fiction* in May and October 1941, respectively. They were combined and are available today as *Orphans of the Sky* [51]. Taken together, they represented the first novel-length treatment of the multigenerational starship idea.

The action took place on board a giant interstellar vessel whose crewmembers had forgotten the purpose and nature of their mission as a result of a long-ago mutiny. Most of the inhabitants had become illiterate farmers who had devolved into a pre-technological culture marked by superstition and a belief that the ship was the entire universe. They were led by a ruling oligarchy, and they had to defend themselves against a group of mutants who occupied the upper decks of the giant ship. Through his interaction with the mutants, the protagonist discovered the old control room and learned the secret of the interstellar mission. Together with the two-headed mutant leader, he rebelled against the oligarchy and ultimately succeeded in escaping with a few fellow crewmembers to colonize a new planet. Throughout the story, Heinlein concerned himself with the plight and emotions of the crewmembers (farmers, mutants, and leaders alike) and to issues related to the sociology of the culture. He also described features of the multigenerational ship in memorable and vivid terms (e.g., putting trash and rebellious people into an energy converter to produce the ship's power), and he revealed important aspects of the plot through the eyes of the characters rather than relying on info-dump statements.

In the era that Caroti terms The Birth of the Space Age (1946–1957), the notion of technology producing a better future was combined with Cold War fears of a potentially dangerous present. The number of starship stories began to increase, and some of the trends of the previous era were included and elaborated upon (e.g., better descriptions of technological features; concerns with the psychological well-being of the inhabitants; attention to the ecology of the environment, such as the use of hydroponic systems for food supply).

An example was Frank M. Robinson's "The Oceans are Wide," a novella that was first published in Science Stories in April 1954 and appears today in the collection of related stories simply called Starships [4]. In this story, a Predict who was made immortal by a longevity serum subtly guided the affairs of a multigenerational starship populace. The overt governance was by a hereditary board of executives whose chairman was dying and whose son was ill-equipped to take over due to his own personality and to the scheming of competitive family members. The ship was the only survivor of a fleet that left the Earth 500 years previously following a nuclear holocaust. With the help of the Machiavellian-like Predict, the son grew up to become the new chairman, and the ship successfully reached a distant planet that was geographically and meteorologically idyllic. Fearing that such a place would retard the development of his followers due to the lack of challenges to overcome, the protagonist drugged his people and continued on to another planet that had a much bleaker and stressful environment. He felt that this planet would stimulate the colonists to advance scientifically and politically as they overcame adversity. Outwitting his teacher, the protagonist himself became the new Predict, showing that he learned his lessons very well indeed! In this story, the nuclear fears of the Cold War were front and center in providing the rationale for the mission, as were ecological concerns and the need to create a better society based on conquering new challenges. Life on board the multigenerational ship was clearly described, including recreational activities, the intrigues of governance, and the need for population control by euthanizing people at 60 years of age. The story clearly reflected the times.

Things changed dramatically in the New Wave era (1957–1979). The launch of Sputnik, the Viet Nam War, and the rise of feminism and the hippie generation all served to dethrone traditional male American values as the only acceptable blueprint for the future. In addition, science fiction extended into venues beyond pulp magazine stories (e.g., novels, television shows, movies, board games), and it increased its scope and fan base to include more women, minorities, and juveniles. A number of sub-genres emerged, such as feminist or young adult science fiction and science fantasy, and the more traditional hard science fiction itself became a sub-genre. These trends were reflected in interstellar travel stories as well. According to Caroti, some two dozen multi-generational starship narratives appeared during this era, about ten times that of the Gernsbeck period and double the frequency of the Campbell era [21, p. 201]. Strongly influenced by more liberal and socially-motivated British editors and writers (as exemplified in the pages of *New Worlds* magazine), the New Wave put more emphasis on the biological and social sciences than the physical sciences, added Earth-relevant themes to older themes focusing on space technology, and encouraged new writers who had more mainstream literary values and styles.

An example was British writer Brian Aldiss' novel *Non-Stop*, originally published in 1958 [2]. The story started out like a typical forgotten mission multi-generational story, where the protagonist was a member of a primordial tribe living in a primitive environment that turned out to be a gigantic starship overrun by vegetation. After he and some colleagues began exploring their environment, they encountered a more sophisticated tribe and learned that the ship's inhabitants were returning from a planet around the star Procyon, where they had suffered from a pandemic some 23 generations earlier due to an alien amino acid that was present in the water. They also encountered a race of giants and a mutated species of sentient rats. The plot surprise is that they had been orbiting the Earth for some time, and the giants were in fact normal-sized Earthlings trying to cure them of the effects of the pandemic. These effects included a mutation-generated speeding up of their metabolic rate (and sense of time) and a shortening of their stature. By focusing on the unfortunate plight of the original starship travelers, the biological issues related to their environment and their mutations, the use of psychoanalytic theories and terminology to describe the rituals and interactions of the inhabitants, and the gradually revealing twists and turns in the storyline, this novel fit well into New Wave sensibilities.

Caroti's final formal era is the Information Age (1980–2001). Reflecting the growth of computers and smart phones, science fiction once again took on a technological character. However, it was of a more personal, intimate nature and was promulgated by a new group of technology-savvy writers. One example was cyberpunk, which examined the edgy relationship between technology (especially information-based) and the human body. Another approach examined nanotechnology and its impact on people. During this era, multigenerational starship stories, with their large vessels, lofty goals, and faraway destinations, decreased in number. Caroti states that only seven multigenerational starship narratives appeared between 1980 and 2000, without a corresponding loss of science fiction publication venues [21, p. 201].

One story he discussed was Bruce Sterling's 1998 novella "Taklamakan." In this story, the two protagonists used digital camouflage suits, miniaturized biotechnological mission gear (e.g., gelbrain cameras, pre-programmed neural tissue, drills running on sugar enzymes), and information-seeking equipment to locate a Chinese rocket base buried under the desert surface. Here they found three giant space ships inhabited by separate groups of people considered undesirable by their government and imprisoned during their supposed "interstellar mission" in order to isolate them from society. When people from one of the groups tried to escape, they were attacked by a swarm of robots and reconfined. Biomechanical vats were used to replace the depleted ranks of the robots with new improved models that then managed to break through to the surface, revealing the horror of the underground prison to the people above.

In looking at trends over the first decade of the twenty-first century, Caroti states that over a dozen multigenerational starship narratives appeared, reflecting a rise in number over the two decades that ended the preceding century. To explain these increasing numbers, he states that:

> in an age where the exponential acceleration of data exchanges across the globe seems to have commensurately speeded up the pace at which generations succeed one another, the variables portrayed in generation starship narratives— transfer of information and values from parents to children, retention or loss of cultural identity—have perhaps become more important than they have ever been before. [21, p. 243].

It is also possible that long-duration missions onboard the International Space Station and plans for a future manned expedition to Mars have stimulated interest in extended missions in space, especially as we are discovering more and more planets orbiting distant stars (see below). Caroti goes on to say that the genre seems to be revisiting past novums and attempting to bring them up to date by reconfiguring them into "new fictional shapes." Certainly, recent collections of short starship fictional narratives may be seen as attempting to do this [13, 61].

Suspended animation also has been revived in new forms. For example, in Charles Gannon's 2013 *Fire with Fire* [41], the hero was placed in cryogenic sleep for 13 years by a secret government agency for security reasons, and he later escaped an assassination attempt by abandoning a space ship via an escape pod that automatically put him into suspended animation for the trip home. But as we shall discuss below, putting an entire human being into suspended animation is likely to be a difficult procedure, with both psychological and physiological ramifications, and these difficulties need to be addressed further by science fiction authors.

Examples from the Novel The story in this book, *The Protos Mandate*, reflects some of these newer tendencies. The storyline deals with the construction, flight, and arrival of a multigenerational starship to a distant star, finishing with the landing of the crew on an orbiting planet, Protos. There are two driving forces for the mission. The first relates to the effects of global warming and overpopulation on Earth and the need to decompress the situation by moving people out, first into the solar system (which has been done), then to the stars. The second stimulant concerns the ongoing search for life on other worlds beyond those in our solar system. The propulsion system used for the starship, the political and economic issues affecting the mission, the psychological and sociological pressures experienced by the crewmembers along the way, and the vicissitudes related to what they find on Protos reflect projections based on current science and technology. In addition, new findings regarding exoplanets are incorporated into the storyline. But some traditional ideas also are used in new ways. For example, both multigenerational issues and issues related to suspended animation are conjoined in the story. Also, the struggle to promulgate the original cultural values and objectives of the mission through multiple generations is explored. This is contrasted with the need to be flexible enough to alter the original "mandate" to reflect new realities. Finally, *The Protos Mandate* is a human story dealing with people engaged on a new enterprise lasting more than 100 years. Their feelings, motivations, and relationships are examined, in some cases changing as they grow and mature in the world that is their starship.

2 Traveling to the Stars: Distance, Propulsion, Radiation

In considering where to go, the stars closest to us are the likeliest candidates for the first multigenerational starship mission. In our Sun's neighborhood, the closest stars and their distances in light-years (in parentheses) are: Proxima Centauri (4.2), Alpha Centauri A and B (4.4), Barnard's Star (5.9), Wolf 359 (7.8), Lalande 21185 (8.3), Sirius A and B (8.6), UV Ceti A and B (8.7), Ross 154 (9.7), Ross 248 (10.3), and Epsilon Eridani (10.5) [54].

All of these stars are a long way away: trillions of miles. Using current technology, interstellar travel is highly unlikely. For example, a starship traveling at the same speed as Voyager 2 would take around 497,000 years to reach the Sirius star system [107]. In contrast, a ship traveling at 5 % the speed of light (.05c) would take 88 years to reach Alpha Centauri [107]. Although an improvement, this would still be longer than the expected lifetime of most of the crewmembers and would necessitate a multigenerational approach or the use of suspended animation.

Since faster than light speeds, travelling through wormholes, or using a "warp drive" to distort space-time are not scientifically credible options at the present time, new propulsion systems that can reach a significant fraction of the speed of light will be necessary. In a typical mission, the vehicle must first accelerate up to this speed, then coast along through much of the mission at this velocity, and finally decelerate to orbital or landing speed as it approaches its destination. It has been calculated that by accelerating such a starship at the force of one *g* (producing an Earth-like gravity situation for the crewmembers), it would take about a year to reach a cruising speed close to that of light [74, p. 11]. The acceleration time would be less for a ship reaching a more manageable cruising speed, say around 10 % the speed of light (.10c). Relativistic time effects are important close to light speed, but they are relatively negligible at speeds in the range of .10c.

Three kinds of propulsion system have been identified for interstellar missions: those that carry their own fuel, those that rely on some sort of external energy source to move them along, and hybrids of these two.

Interstellar Vehicles Using Internal Energy Sources Traditional rocket-based propulsion systems are self-contained: they carry along their reaction mass, energy source, and engine, all of which greatly increase their total mass and costs. One type is the nuclear fission rocket, which uses a nuclear reactor to thermally accelerate hydrogen atoms to provide thrust; a variant adds a thermal-to-electric generator to expel charged atoms at high velocity (the nuclear electric rocket) [37, 77, 78, 81, 82, 93]. An example of the former was a program called NERVA (Nuclear Energy for Rocket Vehicle Application), which developed some prototype engines in the late 1950s and 1960s but was terminated in the early 1970s. An example of the latter proposed by the Jet Propulsion Laboratory in California in the mid-1970s was the TAU (Thousand Astronomical Unit) mission. Although useful for outer solar system travel and transport, such fission rockets do not produce enough thrust to reach a star in a reasonable amount of time.

A second and more powerful system that contains its own energy source is the nuclear pulse rocket, which is propelled by small nuclear bombs ejected and exploded every few seconds or so against a heavy-duty pusher plate at the back [29, 37, 76, 78, 81, 82, 84, 93]. The pusher plate absorbs each impulse from the hot plasma and transfers it to the vehicle through large shock absorbers. The prototype system for this method of propulsion was Project Orion, proposed by the Los Alamos National Laboratory in New Mexico in the late 1950s and early 1960s to use small nuclear fission bombs and in the late 1960s by Freeman Dyson to use fusion devices [29, 78, 82, 84]. It was estimated that some 300,000 bombs would be needed to propel the massive space ship, which would weigh 400,000 tons and accommodate a crew of several hundred [37].

A third internal energy system was explored by members of the British Interplanetary Society in the 1970s and was termed Project Daedalus [12, 29, 39, 76, 78, 81, 82, 84, 92, 93]. This was a fusion-powered interstellar rocket where pellets of helium-3 and deuterium were compressed and heated in a combustion chamber inside the ship by high-energy electron beams or lasers. The resulting fusion reaction provided energy to power the vessel. Although some have proposed using tritium instead of helium-3 since the reaction is easier to initiate, helium-3 results in charged particles that can be confined and directed by a magnetic nozzle (rather than the leaky neutral neutrons that are produced by the tritium reaction) [76, 82, 84, 92]. Since helium-3 is rare on Earth, it would have to be mined elsewhere, such as the atmosphere of Jupiter or Saturn, possibly using robotic helium mines suspended by balloons [82, 84]. Deuterium could be obtained from cometary nuclei in the Oort cloud [122]. A follow-up version to Deadalus, Project Icarus, was examined in 2009 to explore similar concepts using newer twenty-first century notions and to develop some sort of deceleration mechanism when the target star was reached [12, 79, 84, 92, 95]. Other variants included the novel use of autonomous robotics and artificial intelligence for on-board planning, maintenance, and self-repair [36], and new propulsion concepts, such as plasma jet driven magneto-inertial fusion [110]. Another Daedalus follow-up examined the use of the entire spacecraft as a magnetically-insulated capacitor which would ignite the deuterium-tritium reaction using an intense ion beam [123].

A fourth internal energy system depends upon the reaction of matter and antimatter to provide energy to move the vehicle [37, 78, 81–83, 93]. Although the concept has been discussed since the 1950s, it was more fully developed in the early 1980s by Robert Forward [37]. The notion was that the reaction of protons and antiprotons would produce electrically charged elementary particles that could be focused by a magnetic nozzle and expelled out the back of the rocket ship as exhaust. Although more powerful than fission or fusion, this system presents technical issues related to storing antimatter in a manner that would prevent it from touching and reacting with the walls of the ship, such as in a magnetic or electric field. In addition, antimatter is very rare, and it would be a challenge to obtain enough of it to propel a giant starship.

Hybrid Interstellar Propulsion Systems External energy propulsion systems solve the major problem that decreases the efficiency of systems using internal energy: the need to take along large amounts of heavy fuel. Hybrid systems likewise rely on external energy sources to decrease mass, but they also use small amounts of internal energy. One such system was the Bussard interstellar ramjet, which was proposed in the early 1960s by Robert Bussard [29, 37, 78, 81, 82, 84, 104]. This vehicle consisted of the payload, a fusion reactor, and a large electrical or magnetic scoop to collect onrushing charged particles along the flight path. Interstellar hydrogen was the main fuel source. However, some supplemental intrinsic fuel was necessary for travel through low hydrogen areas, such as in our Sun's vicinity [29, 74]. Although this model employed a heavy rocket engine whereby the energized helium exhaust resulting from hydrogen fusion was expelled from the rear of the spacecraft to accelerate it forward, such a starship would not need to carry a lot of fuel during the trip, thus cutting down on mass and cost. Since the amount of hydrogen collected by the ramscoop increases with speed, this system could reach high velocities and would be suitable for interstellar travel, assuming it was designed well enough to minimize drag. The scoop would need to be large and structured using lightweight material, or it could consist of a magnetic or electrostatic field that would collect hydrogen that has been ionized by a forward pointing laser [82].

One variant of the Bussard approach is the Ram-augmented Interstellar Rocket (RAIR) [81, 82]. This system incorporates a separate fusion reaction that uses a small amount of intrinsic fuel such as helium-3 and deuterium (see above). But in this case, the reaction serves to energize the hydrogen that is collected from space by a ramscoop, which it does in a very efficient manner. Note that the hydrogen is not used as fuel but as reaction mass to produce thrust for the starship.

Interstellar Vehicles Using External Energy Sources A purely external energy system discussed as far back as the 1920s employed beamed power. The type usually mentioned uses the momentum of massless light photons from the Sun to "push" against a solar sail, thus moving the vehicle in the direction of the beam [11, 12, 29, 37, 60, 62, 76, 78, 81, 82, 93, 104]. In contrast to the solar-electric drive that uses sunlight falling on solar cells to convert fuel to ion propulsion [21], a beamed system would only need a payload and the structure of the vehicle; there would be no need for heavy intrinsic fuel or any kind of engine. The solar sail concept has been tested on Earth and in space with some success by several space agencies [40, 60, 82, 85]. Being located within our solar system, the beaming system could be monitored and maintained relatively close to home. However, the space vehicle would be a relatively slowly accelerating system, and the larger the payload, the greater the need for a very large sail. This system would likely be better for unmanned interstellar missions carrying small payloads.

A number of other beam/sail systems have been suggested, such as using small charged pellets accelerated by an electromagnetic mass driver which strike a magnetic field sail; microwave photons pushing against a wire mesh sail containing microcircuits at the wire intersections; or lasers aimed by a Fresnel lens reflecting against a large light sail [11, 12, 29, 37, 62, 76, 81, 82, 104]. Much like a tacking sailboat, some of these systems allow the craft to turn or even decelerate upon reaching a stellar destination. Methods for using the solar wind have also been considered for travel in the solar system [58, 59, 99]. More novel approaches for beamed propulsion have been proposed as well, such as using gravitational waves [82, 88] and antimatter [72] to generate thrust.

****

Several of the above propulsion systems are capable of achieving very high speeds that would cut down on travel time. A round trip to Proxima Centauri could be made in 11 years, assuming a 1-year acceleration to near-light speed, then a 3½-year coast in deep space, a 1-year deceleration to the star, then a similar flight plan on the return [15]. But traveling at near-light speed presents difficult technological problems. In addition, the rapidly oncoming flow of interstellar gas and dust particles and cosmic rays striking the starship and its inhabitants could present unique particulate and radiation hazards [105]. Some kind of deflector shield and laser combination in the front will be necessary to block oncoming dust particles and vaporize larger bodies, although it has been pointed out that a massive deflector system might interfere with the maneuverability of a starship traveling at relativistic speeds [106]. In a ramjet type of vehicle, micron-sized bits of dust likely will be vaporized by protons in the electromagnetic field of the scoop. To protect against oncoming cosmic rays, a passive rock or metal shield or an active magnetic or electric field deflector could be used [81].

Examples from the Novel In *The Protos Mandate*, small unmanned probes using beamed technology are sent to a number of close stars to look for habitable planets for colonization to relieve the population pressures on Earth and in our Solar System. After results are reported back suggesting that Epsilon Eridani is a good candidate to harbor such a planet, a massive fusion powered multigenerational starship is built using hybrid principles based on the Ram-augmented type of Bussard interstellar ramjet, with a ramscoop designed to minimize drag as it collects interstellar hydrogen. This ship is capable of reaching a cruising speed of 10 % the speed of light and creating an internal gravity of 1-*g* during its periods of acceleration and deceleration. During the cruise phase of the mission, the crew is able to work under 1-*g* conditions as a result of their being situated in a giant habitation wheel revolving around the central core of the starship.

3 Economic Considerations

The technology to propel and protect a starship would be enormously complicated and expensive, especially when one considers the massive size of the ship itself. Consider the scenario of a huge, self-contained multigenerational starship full of colonists needing to be kept alive for decades while traveling to a distant star. Strong [111] has envisioned giant 100-megaton starships containing 100–150 people that would be equipped for a century-long journey to the stars. Woodcock [124] imagines even larger one million metric ton starships the length of 11 football fields that would carry 10,000 people. Accelerating to a maximum velocity of 15 % the speed of light (.15c), then decelerating to reach a star some 10 light-years away, such a behemoth would complete its journey in about 130 years.

Zubrin has taken a look at the economics of a starship with a dry mass of 1000 tons that can cruise at .10c and carry a few score colonists on a trip lasting several decades [126]. He estimates that if this ship operates at 100 % efficiency (an unlikely occurrence), the energy costs alone would amount to $ 12.5 trillion. The addition of other costs, such as technology development and hardware manufacture, raises the price tag to $ 125 trillion! This is roughly 1000 times the cost of the Apollo program in today's dollars. He estimates that to keep the cost of this interstellar mission at Apollo levels in proportion to the total wealth of human society (about 1 % of GDP), a future spacefaring civilization will need a GDP 200 times greater than today and a total human population of some 40 billion to make this possible. He foresees fusion using helium-3 and deuterium for fuel as the power source to most cheaply meet the high-power needs of this civilization. The fuel could be mined from the atmospheres of the outer gas giant planets in our Solar System. He believes that the helium-3/deuterium fusion reaction would be the power source for an interstellar vehicle as well, with the super-hot plasma charged particles being confined and reacting in a vacuum chamber using magnetic fields, and the exhaust mixture being directed away by a magnetic nozzle to provide the thrust. A number of technological issues need to be addressed before such a system is possible (e.g., containing the super-hot plasma, using catalytic methods to enhance fusion at a lower temperature), but Zubrin presents a good case. Of course, political, scientific, and economic stakeholder considerations (e.g., national policy priorities, scientific benefits, profit generation) will also influence the likelihood of such a mission [22, 50].

Assuming that there is a will to undertake such an interstellar mission, that appropriate resources are devoted to it in a sustained manner, and that technological breakthroughs occur in a timely sequence, it is reasonable to assume that small, unmanned beam-powered interstellar probes could be launched to nearby stars like Alpha Centauri in the twenty-second and twenty-third centuries [114]. Such probes might even use nanotechnology [82]. After they report back their findings, massive, manned, fusion-powered colony ships could be built and launched in the twenty-fourth and twenty-fifth centuries [80, 86]. Due to the scale and economics of the situation, a fusion propulsion system may not be used for the colony ship. Instead, beamed propulsion might be adequate, especially if several probes are launched sequentially that can use the same beaming source. However, travel by this method would be slow and require much more time to reach the destination stars.

Examples from the Novel In the novel, economics play a key role, since the cost of the Protos mission is in the trillions of dollars, and there is great competition for this money. For example, China wants to relocate billions of its people to more environmentally comfortable regions of its state, and Mars wants to complete its terraforming activities and encourage people to migrate to the Red Planet. In addition, the building of *Protos 1* has turned out to be a major technological undertaking that has demanded major political and economic commitment for nearly three decades, especially given changing conditions and competing priorities in the Solar System. All this is occurring against the backdrop of a struggling Earth that continually siphons off financial and material resources to fight off the effects of global warming and overpopulation, thus slowing down the pace of scientific and technological advances (e.g., the development of suspended animation). Complicating the picture are special interest groups and terrorist organizations that don't want the mission to succeed and attempt to sabotage it.

4 Psychological and Sociological Issues

Most of what we know about psychological and sociological issues in space comes from anecdotal reports from returning crewmembers; scientific studies performed in space analog environments on Earth, such as in the Antarctic or space simulation chambers; and experiments conducted on-orbit [67, 68]. Stressful issues have included: psychophysiological changes brought about by microgravity; psychological sequelae of long-term isolation and confinement; potential danger from radiation or micrometeoroid impacts; extreme monotony alternating with periods of high activity; low cohesion resulting from crew heterogeneity due to personality, gender, cultural, and language differences; and poor or inappropriate leadership. In two large studies of astronauts and cosmonauts on the Mir and the International Space Stations, Kanas and his colleagues found that crewmembers displaced on-board tension outward to mission control personnel; crew cohesion was related to support from the commander; and there were cultural differences reported between Americans and Russians in language flexibility, perceived tension, and work pressure [14, 69]. Other investigators similarly have found evidence for crew tension and its outward displacement, as well as negative effects on cohesion due to sub-grouping and scapegoating [46, 47, 60]. Interpersonal stressors have been reported in astronaut diaries to influence space crew performance [112], and surveys have revealed crew tensions due to cultural differences [91, 102, 116].

Although the above findings have relevance for future near-Earth missions, caution must be used in extrapolating them to deep space missions, especially as regards the colonization of a planet orbiting a distant star. Multigenerational interstellar missions lasting 100 years or more present a number of new psychological and sociological problems for the travelers [63]. These are summarized in Table 1.

First and foremost is the selection of the crewmember colonists. Who would volunteer to undergo such a mission, where family and friends are left behind on Earth, never to be seen again? How will the crewmember selection be made: lottery, scientific value, genetic make-up, politics? Which family members would be able to go along and which would be excluded? How would optimal diversity be established in terms of gender, nationality, religion, occupation, intelligence, etc.? These are all important issues that may affect crew cohesion and the ultimate success of the mission, particularly for the very first expedition where there will be a real sense of adventure and little will be known about what the crew experience will be like.

Table 1.   Psychological and sociological issues during an interstellar mission

1. Selection issues: Who would want to go? Who would be excluded? What kind of diversity would there be in the crew?

2. Feelings of isolation and loneliness in deep space

3. Earth as an insignificant dot in the heavens—Earth-out-of-view phenomenon

4. Lack of novelty and social contacts in deep space

5. Dealing with monotony and leisure time through meaningful activities and habitability design

6. Autonomy from Earth and over-dependence on on-board resources: computers, machinery

7. Dealing with mentally or medically ill people in a confined space

8. Unknown physical and psychological effects of radiation due to traveling at near-relativistic speeds

9. Starship environment: sustainable resources, gravity conditions, population control

10. Intolerance of diversity: cultural factors, religion, language differences

11. Feelings of homesickness, especially people in the first generation who directly remember the Earth

12. Dealing with myths and folklore regarding the Earth in later generations

13. Keeping the original colonizing goals: rebellion by later generations who want to go back or keep traveling in space, flexible governance

14. Dealing with criminals and sociopaths in a relatively small social network

15. Psychological and ethical effects of social engineering: regulating coupling, birth rate

16. Psychological and medical issues related to suspended animation

Once underway, the crew of a starship will experience a severe sense of isolation and separation from the Earth, which will become a distant dot in the heavens, and later just a memory and the subject of history and folklore. What will be the psychological impact of seeing the disappearing home planet, the so-called "Earth-out-of-view phenomenon" [68]? Will this result in profound feelings of loneliness and depression? Direct human contact will be limited to just the crewmembers, and ennui may result from the lack of novelty and the predictability of interacting with the same people for years. This will make work and leisure time activities important to counteract monotony, and careful attention should be paid to having a stimulating habitability design [97]. Starship crewmembers will become completely autonomous from the Earth. Long delays will occur in communicating with home, although this communication may be enhanced by novel approaches, such as using the Sun as a gravitational lens to improve communication signal gain [42, 43]. Crewmembers will have to depend on themselves and on-board computers and machinery for basic life support and operational issues, which may present psychological concerns about hardware and software reliability. The crew will also need to deal directly with all problems and emergencies, such as might be produced by a psychiatrically-disturbed crewmember or by a medical problem that affects general crew health. In addition, the physical and psychological effects of radiation may become a factor over time in deep space. This impact may be enhanced as the ship accelerates to a significant proportion of light speed, thus producing an oncoming intense flow of interstellar particles and cosmic rays [105], much as what happens when driving in a snowstorm on Earth. Work is underway to assess cosmic ray shielding issues using computer models [125] and technologies related to physical or electromagnetic shield systems [24, 81].

There are a number of issues that will arise from the unique environment of the starship that will affect crewmember well-being. Equipment and supplies must be sustainable and recyclable, and the crew will need to be able to maintain and even alter the environment to meet changing needs [3, 24]. Water, food, and medications will need to be produced, and advanced technologies using hydroponics, light, genetic engineering, and even smart microfarms employing algae should be explored for possible implementation [35]. For such a long multigenerational mission, artificial gravity approximating that of Earth will be necessary. This could be produced by spinning the ship along its longitudinal axis (provided it is wide enough so that Coriolis forces would be minimized) or housing the crew in a wheel habitat revolving around the central core of the starship [24]. The birthrate will need to be controlled in accordance with existing resources to account for unexpected deaths due to illness, epidemics or starship accidents.

Sociologically, issues related to the crew composition will be important. How much cultural and religious diversity will be tolerated? Will people in the minority be scapegoated during stressful times? Will subgroups emerge that would impact on crew cohesion? Although a common language is likely to be used, would alternatives be acceptable? What about changes that occur from one generation to another? For example, there likely will be new vocabulary words added that are relevant to an interstellar mission, and some "Earth-based" words that are irrelevant might disappear (such as "snow" or "boat") [115]. What kind of social structure will be optimal to deal with diversity given the limited space and resources? All of these issues need to be carefully thought out prior to launch. Possibly, the first generation of colonists would be most affected by these issues. In addition, they would vividly remember the Earth, having left friends and associates behind. This could produce strong feelings of homesickness in this group. Subsequent generations would likely be less affected, since their total existence and reference point will be the multigenerational starship itself. Nevertheless, images and stories of the Earth will no doubt be preserved and will likely be the subject of future folklore and longings as time goes on.

Another generational difference might result from the need of the first generation to be true to the original goals and objectives of the mission, whereas later generations might be more flexible. Some might even rebel and want to return to Earth. Alternatively, they might want to avoid landing on the targeted exoplanet, preferring to keep going as a permanent spacefaring people without achieving any landfall. The governance system will likely reflect the goals and objectives established years earlier prior to launch, but it will also need to be flexible enough to accommodate changing conditions during the flight as well as after landing on a distant extrasolar planet [3, 24]. If there is a rebellion, how will the rebels be dealt with? Will there be a jail for criminals and sociopaths? What sort of legal system would there be, and how would it be enforced?

A multigenerational interstellar mission raises important moral and ethical issues [100]. For example, the ethics of applying social engineering principles to keep the population size under control may be questioned. Will men and women pair off into stable couples, or will a more sexually free society evolve? What will be done as regards the practical need for birth control versus the emotional desires of couples to procreate? How will the society accommodate people with strong religious views about going forth and multiplying? When will people be allowed to have children, and how many? How will children be raised: in traditional family units or communally?

Moore has discussed some ideas about population control in a hypothetical multigenerational starship crew of 150–180 people on a 200-year mission to Alpha Centauri [87]. Based on his computer modeling techniques, he has enumerated several social engineering principles that would establish a stable population and maximize the group's productivity. In general, he favors the traditional kinship family organization. However, he would begin the mission with a crew of young, childless married couples who would agree to postpone parenthood until late in a woman's reproductive life, say in her 30s. This would result in smaller sibships that would require less time for childrearing and free up more time to accomplish mission tasks. This society of monogamous couples and small families would help maintain genetic variation by increasing the proportion of unrelated persons in the marriage pool. Over time, well-defined demographic groups would result, with people of roughly the same age clustered into discrete Echelons. The oldest group would be the experts as well as the teachers and babysitters of the youngest group. The middle Echelon would be responsible for day-to-day mission operations and tasks relevant to the maintenance of the society. If the birth rate was controlled to match the death rate, the population would remain stable and could be adjusted upward if an illness or accident killed more people than expected.

Although this system has a number of positive demographic advantages, O'Rourke has reminded us that advanced parental age correlates with a higher frequency of unfortunate genetic problems in offspring, such as Down's syndrome and achondroplasia [96]. He also reminds us that in small closed populations, genetic drift can lead to less heterozygosity (i.e., gene variation at the allele level), which could result in the expression of recessive phenotypic traits that might be harmful to offspring. Consequently, vigorous genetic testing and prenatal medical care are givens in such a society. In addition, it is possible that the citizens will not accept such social engineering, especially those of later generations who were born into a system that was not of their choosing. Will they protest and choose to have more children at an earlier age? One can imagine this becoming an issue of rebellion by teenagers and young adults determined to establish more control over their reproductive lives.

Examples from the Novel In this story, the social structure envisions four Echelons: Learners, Citizens, Elders, and Grandelders, each around 30–35 years apart in age. Furthermore, there are two separate clans, the green-coded Alphas and the blue-coded Betas. Although protected sex is allowed earlier, people are strongly prohibited from pairing off with others in the same clan and having children together before the age of citizenship (30 years). This creates a problem in the story for two young people from the same clan who are in love and want to pair off. They rebel against the Protos Mandate, which outlines the rules for this closed society and in particular dictates the population control policies on the starship. Another problem is the death of so many Sleepers, who were expected to breed with other crewmembers after awakening. The Sleepers were counted upon to provide novel alleles that would enter the gene pool to help counter the effects of genetic drift and decreased heterozygosity in the awake crew population, so this corrective influence would likely be lessened by the unexpected deaths.

5 Suspended Animation

Putting crewmembers in suspended animation has been a well-utilized novum in science fiction as a way of conserving resources and dealing with the long durations inherent in interstellar missions. It has been employed in both written stories (e.g., Don Wilcox's 1940 "The Voyage that Lasted 600 Years," A.E. van Vogt's 1944 "Far Centaurus") and popular movies (e.g., *2001*, *Alien*). In this scenario, after the critical activities involving the launch and the setting of the course for a distant star have been accomplished, the crew would be put in a state where their physiological functions are slowed down until such time as they are near their destination, when they would be "awakened" to perform their landing and exploration duties. This notion proposes the effective cessation of metabolism in the crewmembers due to drugs and/or extreme cold (i.e., cryosleep). Certain key crewmembers could be revived periodically to perform mission critical activities, then go back into suspended animation when these are completed. The starship would be on autopilot during the bulk of the mission, and computers would handle life support and navigation, as well as the revival process.

The problem is that the technology to put an entire human being in suspended animation has yet to be developed, and the process is fraught with difficulties. Although freezing is used to preserve red blood cells and corneas for transplantation, the ability to freeze and later thaw complete organ systems and whole bodies composed of differentiated cells with different freeze-thaw rate profiles is beyond our abilities in the foreseeable future [81, 82, 113]. Ice crystals can form, which can be lethal to cells, and areas of the body can be deprived of oxygen from blood clotting or premature freezing before metabolism is slowed down. Even the use of cryoprotectants such as glycerol, sucrose or ethylene glycol presents technological challenges [113]. The thawing of previously frozen cells and tissues presents risks of ice crystal formation and damage as well.

A related idea is to cryopreserve sperm, ova or actual embryos in liquid nitrogen or via other techniques for later implantation in female crewmembers or in an artificial womb [28, 82, 113]. This would present a possible backup system for fertility problems that might develop in transit to a distant star, or it could be used to increase the colony population after landing on a suitable exoplanet. Such preservation for up to two decades has resulted in successful implantation and birth [113].

One notion of preserving cells in the human body is through the process of vitrification. In this process, the water in the body and its cells is cooled in such a way that it does not actually freeze. Instead, it is supercooled to a kind of glass-like state where cellular molecular motion and metabolism cease and cell components are preserved in place due to the arrested state of motion [113]. Although in theory the dangers of freezing should not be present, ice crystal formation and cell damage could still occur during the thawing process.

But even if suspended animation becomes technically possible, problems could still occur. Perhaps there are unknown physical and physiological effects of long-term suspended animation lasting up to a century or more that might result in permanent organ damage or impaired brain function. This risk could be enhanced by power surges or breakdowns of the equipment during this long period of time. In addition, psychological problems could result prior to freezing in people fearful of being incapacitated for years at a time or worrying that some threat could occur, such as a collision or equipment failure. For example, what if a meteoroid hit the ship and impacted negatively on life support equipment before crewmembers could be aroused? Computers and other machines are not perfect; the notion of being helplessly dependent on them to maintain your life and revive you later on is not a comfortable thought and could create anxiety. Many people would prefer the awake multigenerational option for the first space colony mission, since they would be in more control over their destiny.

Examples from the Novel In this story, 40 Sleepers have volunteered to undergo suspended animation. These people have been trained specifically to function in a colony located on a new world. Since their skills would not be useful in transit, and since if fully awake their metabolic needs would use up limited food, water, and oxygen, then having them function as part of the awake crew would not be a good use of limited resources. However, the suspended animation procedure carries some risk, especially since it had not been tested in people for longer than a few years. For this reason, most volunteers for the Protos mission have opted for the awake option. Indeed, in the story more than a third of the Sleepers could not be revived when Protos was reached, and this created problems for the remaining crewmembers (e.g., loss of skills needed for the new settlement on Protos, less contribution of novel genetic alleles for the colony gene pool).

6 Exoplanets and Colonization

Planets revolving around distant stars can be detected using several techniques, such as astrometry, which measures a star's wobble due to the gravitational influences of an orbiting planet; Doppler changes in stellar spectrum due to this wobble; pulsar timing variations resulting from planet-caused gravitational perturbations as the pulsar rotates; changes in a star's luminosity resulting from a transiting planet; and gravitational microlensing, where the light from a background star is bent by the gravitational effects of a closer in-line star with planets [25, 65, 81, 82]. Often, the mass and distance of the exoplanet from its star can be determined. These detection methods bias the search in favor of finding larger planets, but as the techniques become more refined, more and more exoplanets approaching the size of Earth are being discovered. Thanks to the sensitivity of the Kepler Space Telescope, the NASA Exoplanet

Archive on April 1, 2014, listed 1693 planets orbiting 1024 stars [55]. More exoplanets continue to be listed every week as the Kepler data are processed. Some of planets are in the star's so-called habitable (or "Goldilock") zone: not too hot or too cold, but at the right distance to have surface temperatures in the range supporting the presence of liquid water, thus making them possible candidates for life. In fact, a recent study found ten Earth-size exoplanets orbiting in their respective star's habitable zone [98]. The study results supported the conclusion that 22 % of Sun-like stars in our galaxy may in fact harbor Earth-size planets that orbit in their habitable zones, and that the nearest such planet may well be within 12 light-years from us. Nineteen single or double star systems lie within this distance [27, 94].

Three of these systems are thought to have at least one planet orbiting a star [82]. In November 2012 an Earth-like star was thought to have been detected around Alpha Centauri B, located 4.4 light-years from Earth. If confirmed, the planet would likely be very close to its star, and therefore too hot to be habitable [9]. Work published in December 2012 has suggested that the Sun-like star Tau Ceti, located 11.9 light-years away, may host a system of up to five planets ranging in size from two to seven Earth masses, and that two of these are close to the habitable zone [9].

A bit closer to us at 10.5 light-years away, and better studied than the other two star systems, is the interesting system around Epsilon Eridani. With an apparent magnitude of 3.7, this young star is probably less than a billion years old and has a mass of about 80 % that of our Sun [6, 19, 34, 119]. It is of spectral class K2 and has an orange hue. A number of components are thought to surround the star [9, 23, 27, 57, 82, 94, 119]. These include: an inner asteroid belt some 3 astronomical units away (1 AU = the Earth-Sun distance, or 149,597,871 km) [16, 23, 57, 119]; a large planet discovered in the year 2000 that is likely 1.5 times the mass of Jupiter and is around 3.4 AU away from its star, with an orbital period of about 7 years [10, 19, 23, 49, 65, 57, 119]; an outer asteroid belt some 20 AU away [23, 57, 119]; a more Earth-size planet about 10 % the mass of Jupiter around 40 AU away, with an orbital period of some 280 years [23, 119]; and a Kuiper belt-like dust disk 35–90 AU away that is relatively devoid of cometary nuclei [5, 23, 26, 45, 57, 119]. There is speculation that other planets exist in the system, especially bordering and helping to form the belts and disk.

Young K2 stars like Epsilon Eridani are seen as good possibilities to harbor planets that support life. This is because they are numerous in number, are stable for long periods of time, and potential planets orbiting them are less likely to be trapped in a synchronous rotation due to tidal damping than planets around older stars [71, 74, 119, 120]. Although determining the location of a star's habitable zone is dependent upon many factors, such as the star's age, luminosity, and flare activity, as well as assumptions about a planet's magnetic field, climatic conditions, and cloud formation [44, 71, 73, 75, 117, 119, 120], a reasonable estimate of the distance of the habitable zone of Epsilon Eridani is around 0.5–1 AU [108, 119, 120]. Furthermore, with a distance of around .5–.6 AU from this star matching the solar constant and UV flux experienced on Earth, this distance looks promising for any planet found in this location to harbor life [17, 73, 119]. Recently, the Kepler telescope discovered two Earth-size planets orbiting another K2 star (Kepler-62) that is two-thirds the size of our Sun and is located 1200 light-years away from us in the constellation of Lyra [90]. No Earth-size planets have been found yet in the habitable zone of Epsilon Eridani, but should they exist, this would be a good place to look for extra-solar life.

In time, it is likely that exoplanets will be found relatively close to us that are good candidates for colonization. If so, what would such a colony be like? Based on his analyses of 13 post-migration communities on Earth, Schwartz has conceptualized three typical stages of organization following a migration [103]. The first is the pioneering phase, lasting 2–4 years, where the new settlement may experience tension and factionalism over issues related to physical survival. After food has been provided in a reliable manner, and after permanent shelters have been established, this sense of impermanence disappears. The community now enters into the consolidation phase, where it crystallizes and formalizes its social institutions and associations, and a sense of group solidarity begins to develop. In some colonies, there is pressure to retain the old ways of doing things despite changing conditions, but in others new norms are established and cultural changes occur. As the potential factionalism of the first two stages are dealt with, and ways of resolving disagreements are established, the community enters into the third phase, stabilization, where it continues to develop in ways not directly related to the resettlement. Although initially the settlers may experience a sense of equality with each other, the social class structure of the original migrating group could be reestablished later on. Alternatively, new social interactions may result from the new conditions. In a similar manner, either weak or strong authority systems could occur, largely as a result of the nature of the structure in the pre-settlement culture. In terms of religion, Schwartz outlines three patterns: a simplification of the religious system in the early years following the migration; a rise in its importance as a factor increasing the unity of the community; or as a vehicle for factionalism after the initial period of settlement [103]. How these factors will apply to a new interstellar community is dependent upon the specific conditions and social conventions of the group. Economically, Hodges has written that a newly settled star system community will experience a period of great scarcity of goods, but after basic survival needs are met, and after the population has grown and becomes self-sufficient, the standard of living will improve as industries are established that produce goods beyond the basic necessities [52].

Examples from the Novel In *The Protos Mandate*, the starship crewmembers find a planet orbiting Epsilon Eridani that has a number of Earth-like conditions, as they expected from the findings of an earlier robotic probe. This allows them to land and live on the planet's surface without special equipment, like a space suit or pressurized habitats, even though the oxygen level is a bit low, and the weather is a bit damp. In setting up their colony, a variety of social challenges and tensions need to be dealt with, which is typical in the early stage of a new settlement. For the Protos colony, these include making allowances for individual differences and philosophies and dealing with people who are hostile to the new settlement. Nevertheless, the colony seems to be moving forward, with the formation of a democratic governance system and food and water systems in place. However, a serious threat unexpectedly arises from a seemingly innocent source.

7 Extraterrestrial Life

Could life evolve on a planet orbiting a distant star, especially one like Epsilon Eridani that is less than a billion years old? On Earth, there is fossil evidence that suggests that primitive microbes had developed in shallow ocean environments by 1 billion years, and that these organisms evolved in many ways, from obtaining their energy through chemical means (chemoautotrophs) to using photosynthesis (photoautotrophs) [48, 57]. There likely was little oxygen in the atmosphere at this time, but later on the increasingly wider use of photosynthesis began to change things, as atmospheric carbon dioxide was consumed and oxygen was produced. Irwin and Schulze-Makuch have provided intriguing arguments that under the right conditions, the life evolutionary process can be speeded up as compared to that which took place on Earth, and that such a process could have happened on Mars [57]. Specifically, they believe that a billion years would be long enough for multicellular aquatic plants and colonial filter feeders to develop in water environments, and for unicellular extremophiles and organisms living in rock crevices to develop in subterranean and surface environments. With this amount of activity, it is possible that oxygen would have accumulated relatively early in the atmosphere as a byproduct of ongoing photosynthesis. It is unclear how likely photosynthesis would be in the light of a low-luminosity K2 star like Epsilon Eridani. But it should be kept in mind that 4.4 billion years ago, shortly after the Earth was formed, the Sun's brightness was 25–30 % less than today, and that its relative faintness continued for at least another 1.5 billion years [48]. Even under these conditions, photosynthesis-using plants managed to develop and eventually produce oxygen that forms the basis for our existence.

Irwin and Schulze-Makuch further speculate that life could be present in such exotic environments as a watery subsurface on Europa or in aqueous ammonia or liquid ethane habitats on Titan [57]. Alien life has been depicted in a variety of ways living under a variety of conditions [8], but an exoplanet that has been carefully selected for human colonization will likely have a number of Earth-like characteristics with respect to gravity, a rocky surface, moderate temperatures, tolerable radiation, an atmosphere with oxygen, liquid water, and plant-producing soil [31, 53]. As a result, any life found will likely be carbon-based and require sunlight and water. But even on Earth there are a number of extremophilic microorganisms that survive under inhospitable conditions of temperature, radiation, acidity/alkalinity, and pressure [31], and some give off methane as a metabolic byproduct [66]. Organisms with silicon-based structures exist, and there is evidence that silicon may have played a role in the emergence of life on Earth [20, 32]. So it is anybody's guess as to what kinds of alien life future colonists will have to deal with.

One possibility is a life form similar to slime molds on Earth, which are very interesting organisms [18, 33, 118, 121]. Some types live as a syncytium of numerous cell nuclei embedded in a glob of cytoplasm surrounded by a single large membrane. Other types typically exist as single-celled microorganisms that lead solitary lives when their bacterial, yeast, or fungal food is plentiful. However, when food is scarce, they merge together via chemical communication to form a giant amoeba-like organism that is a very efficient finder of food. In studies where the merged organism is placed on a grid depicting a city like London or Tokyo with its surrounding suburbs, and where food is placed at these suburban locations, the slime mold will extend its pseudopods to find the most direct routes to the food, essentially replicating the city's efficient highway or railway system [1, 38]. Similarly, slime molds are able to traverse complex mazes in order to find food [89] and to learn ways of anticipating unpleasant cold and dry conditions in the laboratory [7]. This has given rise to the notion that these primitive organisms possess a kind of rudimentary intelligence [7, 18, 38, 89]. In addition, in their merged state they adaptively form stalks that produce fruiting bodies that release countless spores to reproduce themselves during difficult times [33, 118, 121].

Examples from the Novel In the novel, two life forms are discovered on Protos: algae-like organisms that have adapted to the moist land conditions near the colony, and the "globs" that feed on the algae and are similar to the slime molds found on Earth. The globs are relatively quiescent until the climate warms up and the colonists destroy some of the algae that are their food supply. Stressed out, the globs form larger masses that head for the colonists' own algae food supply. They also produce stalks with organs at the top that release spores. Although the purpose of these spores is reproduction, they happen to be toxic allergens for the colonists. Thus, the settlement is threatened in two ways: loss of food supply, and death from exposure to the spores. The colonists ultimately figure out a way of neutralizing the threat for the time being, but the future is unclear unless a rapprochement is reached between the cohabitating species: human and non-human.

References

1. Adamatzky, A., Jones, J.: Road planning with slime mould: if *Physarum* built motorways it would route M6/M74 through Newcastle. Int. J. Bifurcation Chaos (2009). arXiv:0912.3967
2. Aldiss, B.: Non-stop. Overlook, Woodstock (2005)
3. Ashworth, S.: The emergence of the worldship (II): a development scenario. J. Brit. Interplanet. Soc. **65**, 155–175 (2012)
4. Asimov, I., Greenberg, M.H., Waugh, C.G.: Starships. Fawcett Crest/Ballantine, New York (1983)
5. Backman, D., et al.: Epsilon Eridani's planetary debris disk: structure and dynamics based on Spitzer and CSO observations. Astrophys. J. **690**(2), 1522–1538 (2008). doi:10.1088/0004-637X/690/2/1522
6. Baines, E.K., Armstrong, J.T.: Confirming fundamental parameters of the exoplanet host star epsilon Eridani using the Navy Optical Interferometer. Astrophys. J. arXiv:1112.0447 (2011)
7. Barone, J: #71: Slime molds show surprising degree of intelligence. Discover Magazine, January 2009. http://discovermagazine.com.2009/jan/071
8. Baxter, S.: Imagining the alien: the portrayal of extraterrestrial intelligence and SETI in science fiction. Brit. Interplanet. Soc. **62**, 131–138 (2009)
9. Baxter, S., Crawford, I.: Starship destinations. In: Benford, J., Benford, G. (eds.) Starship Century: Toward the Grandest Horizon, pp. 225–237. Microwave Sciences/Lucky Bat, Charleston (2013)
10. Benedict, G.F., et.al: The extrasolar planet e Eridani b: orbit and mass. Astronom. J. **132**(5), 2206–2218 (2006). doi:10.1086/508323
11. Benford, J.: Sailships. In: Benford, J., Benford, G. (eds.) Starship Century: Toward the Grandest Horizon, pp. 193–204. Microwave Sciences/Lucky Bat, Charleston (2013)
12. Benford, J., Benford, G.: Starships: reaching for the highest bar. In: Benford, J., Benford, G. (eds.) Starship Century: Toward the Grandest Horizon, pp. 1–6. Microwave Sciences/Lucky Bat, Charleston (2013)

13. Benford, J., Benford, G. (eds.) Starship Century: Toward the Grandest Horizon. Microwave Sciences/Lucky Bat, Charleston (2013)

14. Boyd, J.E., Kanas, N.A., Salnitskiy, V.P., Gushin, V.I., Saylor, S.A., Weiss, D.S., Marmar, C.R.: Cultural differences in crewmembers and mission control personnel during two space station programs. Aviat. Space Environ. Med. **80**, 1–9 (2009)

15. Bracewell, R.N.: The Galactic Club: Intelligent Life in Outer Space. San Francisco Book Company, San Francisco (1976)

16. Brogi, M., Marzari, F., Paolicchi, P.: Dynamical stability of the inner belt around Epsilon Eiridani. Astron. Astrophys. **499**(2), L13–L16 (2009). (http://adsabs.harvard.edu/abs/2009A & A...499L...13B)

17. Buccino, A.P., Mauas, P.J.D., Lemarchand, G.A.: UV radiation in different stellar systems. In: Norris R., Stootman, F. (eds.) Structure Bioastronomy 2002: Life Among the Stars. In: Proceedings of IAU Symposium #213, Bioastronomy 2002: Life Among the Stars, Astronomical Society of the Pacific, San Francisco (2003). (http://adsabs.harvard.edu/abs/ 2004IAUS...213...97B)

18. Burton, R.A.: A Sceptic's Guide to the Mind. St. Martin's, New York (2013)

19. Butler, R.P. et al.: Catalog of nearby exoplanets. Astrophys. J. **646**, 505–522 (2006). doi:10.1086/504701

20. Cairns-Smith, A.G.: Seven Clues to the Origin of Life. Cambridge University Press, Cambridge (1991)

21. Caroti, S.: The Generation Starship in Science Fiction: A Critical History, 1934–2001. McFarland, Jefferson (2011)

22. Ceyssens, F., Driesen, M., Wouters, K.: On the organization of world ships and other gigascale interstellar space exploration projects. J. Brit. Interplanet. Soc. **65**, 134–139 (2012)

23. Clavin, W.: Closest planetary system hosts two asteroid belts. http://www.nasa.gov/mission_pages/spitzer/news/spitzer-20081027.html (2008). Accessed 31 Dec 2013

24. Cohen, M.M., Becker, R.E., O'Donnell, D.J., Brody, A.R.: Interstellar sweat equity. J. Brit. Interplanet. Soc. **66**, 110–124 (2013)

25. Coughlin, J.L.: Extrasolar planets: what can be known before going there. J.Brit. Interplanet. Soc. **66**, 47–50 (2013)

26. Coulson, I.M., Dent, W.R.F., Greaves, J.S.: The absence of CO from the dust peak around e Eri. Monthly notices of the Roy. Astronom. Soc. **348**(3), L39–L42 (2004). doi:10.1111/j.1365-2966.2004.07563.x

27. Crawford, I.A.: Project Icarus: astronomical considerations relating to the choice of target star. J. Brit. Interplanet. Soc. **63**, 419–425 (2010)

28. Crowl, A., Hunt, J., Hein, A.M.: Embryo space colonization to overcome the interstellar time/distance bottleneck. J. Brit. Interplanet. Soc. **65**, 283–285 (2012)

29. Crowl, A.: Starship pioneers. In: Benford, J., Benford, G. (eds.) Starship Century: Toward the Grandest Horizon, pp. 169–191. Microwave Sciences/Lucky Bat Books, Charleston (2013)

30. Csicsery-Ronay, I., Jr.: The Seven Beauties of Science Fiction. Wesleyan University Press, Middletown (2008)

31. Dartnell, L.: Biological constraints on habitability. Astron. Geophys. **52**(1), 25–28 (2011)
32. Dessy, R.: Could silicon be the basis for alien life forms, just as carbon is on Earth? Scientific American, 23 Feb 1998. (http://www.scientificamerican.com/article/cfm?id=could-silicon-be-the-basi & print=true). Accessed 4 Sept 2010
33. Douglas, S.M.: Slime molds. The Connecticut agricultural experiment station. www.ct.gov/caes/lib/caes/documents/…/slime_molds_04-02-08r.pdf (2008). Accessed 10 Jan 2014
34. Drake, J.J., Smith, G.: The fundamental parameters of the chromospherically active K2 dwarf Epsilon Eridani. Astrophys. J., Part 1 **412**(2), 797–809 (1993). doi:10.1086/172962
35. Edwards, M.R.: Sustainable functional foods and medicines support vitality, sex, and longevity for a 100-year starship expedition. J. Brit. Interplanet. Soc. **66**, 125–132 (2013)
36. Ellery, A.: Selective snapshot of state-of-the-art artificial intelligence & robotics with reference to the Icarus starship. J. Brit. Interplanet. Soc. **62**, 427–439 (2009)
37. Forward, R.L.: Ad astra! In: Kondo, Y., Bruhweiler, F.C., Moore, J., Sheffield, C. (eds.) Interstellar Travel And Multi-generation Space Ships, pp. 29–51. Apogee, Burlington (2003)
38. Fountain, H.: Slime mold proves to be a brainy blob. New York Times. http://www.nytimes.com/2010/01/26/science/26obmold.html?adxnnl=1 & adxnnlx=13884… (26 Jan 2010). Accessed 30 Dec 2013
39. French, J.R.: Project Icarus: A review of the Daedalus main propulsion system. J. Brit. Interplanet. Soc. **66**, 248–251 (2013)
40. Friedman, L., Garber, D., Heinsheimer, T.: Evolutionary lightsailing missions for the 100-year starship. J. Brit. Interplanet. Soc. **66**, 252–259 (2013)
41. Gannon, C.E.: Fire with Fire. Baen, Riversdale (2013)
42. Galea, P.: Communication with world ships-building the diasporanet. J. Brit. Interplanet. Soc. **65**, 180–184 (2012)
43. Galea, P., Swinney, R.: Project Icarus: mechanisms for enhancing the stability of gravitationally lensed interstellar communications. J. Brit. Interplanet. Soc. **64**, 24–28 (2011)
44. Gilster, P.: In praise of K-class stars. Centauri dreams. http://www.centauri-dreams.org/?p=9032 (12 Aug 2009). Accessed 7 Jan 2014
45. Greaves, J.S., et al.: A dust ring around Epsilon Eridani: analog to the young solar system. Astrophys. J. **506**(2), L133–L137 (1998). doi:10.1086/311652
46. Gushin, V.I., Zaprisa, N.S., Kolinitchenko, T.B., Efimov, V.A., Smirnova, T.M., Vinokhodova, A.G., Kanas, N.: Content analysis of the crew communication with external communicants under prolonged isolation. Aviat. Space Environ. Med. **68**, 1093–1098 (1997)
47. Gushin, V.I., Efimov, V.A., Smirnova, T.M., Vinokhodova, A.G, Kanas, N.: Subject's perception of the crew interaction dynamics under prolonged isolation. Aviat. Space Environ. Med. **69**, 556–561 (1998)

48. Hazen, R.M.: The Story of Earth: the First 4.5 Billion Years, from Stardust to Living Planet. Penguin, New York (2013)

49. Hatzes, A.P., et al.: Evidence for a long-period planet orbiting e Eridani. Astrophys. J. **544**(2), L145–L148 (2000). doi:10.1086/317319

50. Hein, A.M., Tziolas, A.C., Osborne R.: Project Icarus: stakeholder scenarios for an interstellar exploration program. J. Brit. Interplanet. Soc. **64**, 224–233 (2011)

51. Heinlein, R.A.: Orphans of the Sky. Baen, Riversdale (2001)

52. Hodges, W.A.: The division of labor and interstellar migration: a response to "demographic contours." In: Finney, B.R., Jones, E.M. (eds.) Interstellar Migration and the Human Experience, pp. 134–151. University of California Press, Berkeley (1985)

53. Horner, J.: Which exo-Earths should we search for life? Astron. Geophys. **52**(1), 16–20 (2011)

54. NIAC: http://www.niac.usra.edu/files/discover/Star%20Map%20Outreach%20Project.pdf (2014). Accessed 1 April 2014

55. NASA Exoplanet Archive: http://exoplanetarchive.ipac.caltech.edu/index.html (2014). Accessed 1 April 2014

56. Hubblesite: Hubble zeroes in on nearest known exoplanet. http://hubblesite.org/newscenter/newsdesk/archive/releases/2006/32/text (10 Sept 2006). Accessed 31 Dec 2013

57. Irwin, L.N., Schulze-Makuch, D.: Cosmic Biology: How Life Could Evolve on Other Worlds. Springer, New York (2011)

58. Janhunen, P.: Photonic spin control for solar wind electric sail. Acta Astronaut. **83**, 85–90 (2013)

59. Janhunen, P.: Electric sail, photonic sail and deorbiting applications of the freely guided photonic blade. Acta Astronaut. **93**, 410–417 (2014)

60. Johnson, L.: Solar and beamed energy sails. In: Johnson, L., McDevitt, J. (eds.) Going Interstellar, pp. 351–366. Baen, Riversdale (2012)

61. Johnson, L., McDevitt, J. (eds.) Going Interstellar. Baen, Riversdale (2012)

62. Jones, E.M., Finney, B.R.: Fastships and nomads: Two roads to the stars. In: Finney, B.R., Jones, E.M. (eds.) Interstellar Migration and the Human Experience, pp. 88–103. University of California Press, Berkeley (1985)

63. Kanas, N.: From Earth's orbit to the outer planets and beyond: psychological issues in space. Acta Astronaut. **68**, 576–581 (2011)

64. Kanas, N.: Star Maps: History, Artistry, and Cartography, 2nd edn. Springer, New York (2012)

65. Kanas, N.: Solar System Maps: From Antiquity to the Space Age. Springer, New York (2014)

66. Kanas, N.: The New Martians: A Scientific Novel. Springer, New York (2014)

67. Kanas, N., Feddersen, W.: Behavioral, Psychiatric and Sociological Problems of Long-Duration Space Missions. NASA TM X-58067. National Aeronautics and Space Administration Manned Spacecraft Center, Houston (1971)

68. Kanas, N., Manzey, D.: Space Psychology and Psychiatry, 2nd edn. Microcosm/Springer, El Segundo/Dordrecht (2008)

69. Kanas, N.A., Salnitskiy, V.P., Boyd, J.E., Gushin, V.I., Weiss, D.S., Saylor, S.A., Kozerenko, O.P., Marmar, C.R.: Crewmember and mission control personnel interactions during international space station missions. Aviat. Space Environ. Med. **78**, 601–607 (2007)
70. Kanas, N., Sandal, G., Boyd, J.E., Gushin, V.I., Manzey, D., et al.: Psychology and culture during long-duration space missions. Acta Astronaut. **64**, 659–677 (2009)
71. Kasting, J.F., Whitmire, D.P., Reynolds, R.T.: Habitable zones around main sequence stars. Icarus **101**, 108–128 (1993)
72. Keane, R.L., Zhang, W-M.: Beamed core antimatter propulsion: engine design and optimization J. Brit. Interplanet. Soc. **64**, 382–387 (2011)
73. Kitzmann, D., et al.: Clouds in the atmospheres of extrasolar planets. I. Climatic effects of multi-layered clouds for Earth-like planets and implications for habitable zones. Astron. Astrophys. **511**, A66 (2010). doi:10.1051/0004-6361/200913491
74. Kondo, Y.: Interstellar travel and multi-generation space ships: an overview. In: Kondo, Y., Bruhweiler, F.C., Moore, J., Sheffield, C. (eds.) Interstellar Travel and Multi-generation Space Ships, pp. 7–18. Apogee, Burlington (2003)
75. Kopparapu, R.K., et al.: Habitable zones around main-sequence stars: new estimates. Astrophys. J. **765**, 131 (2013). doi:10.1088/0004-637X/765/2/131
76. Landis, G.A.: The ultimate exploration: a review of propulsion concepts for interstellar flight. In: Kondo, Y., Bruhweiler, F.C., Moore, J., Sheffield, C. (eds.) Interstellar Travel and Multi-generation Space Ships, pp. 52–61. Apogee, Burlington (2003)
77. Landis, G.A.: The nuclear rocket: workhorse of the solar system. In: Benford, J., Benford, G. (eds.): Starship Century: Toward the Grandest Horizon, pp. 119–127. Microwave Sciences/Lucky Bat, Charleston (2013)
78. Long, K.F.: Fusion, antimatter & the space drive: charting a path to the stars. J. Brit. Interplanet. Soc. **62**, 89–96 (2009)
79. Long, K.F., Fogg, M., Obousy, R., Tziolas, A., Mann, A., Osborne, R., Prensby, A.: Project Icarus: son of Daedalus-flying closer to another star. J. Brit. Interplanet. Soc. **62**, 403–414 (2009)
80. Long, K.F.: Project Icarus: The first unmanned interstellar mission, robotic expansion & technological growth. J. Brit. Interplanet. Soc. **64**, 107–115 (2011)
81. Mallove, E.F., Matloff, G.L.: The Starflight Handbook: A Pioneer's Guide to Interstellar Travel. Wiley, New York (1989)
82. Matloff, G.L.: Deep Space Probes: To the Outer Solar System and Beyond, 2nd edn. Springer, New York (2005)
83. Matloff, G.: Antimatter starships. In: Johnson, L., McDevitt, J. (eds.) Going Interstellar, pp. 79–101. Baen, Riversdale (2012)
84. Matloff, G.: Fusion starships. In: Johnson, L., McDevitt, J. (eds.) Going Interstellar, pp. 197–217. Baen, Riversdale (2012)
85. Matloff, G.L.: Interstellar light sails. J. Brit. Interplanet. Soc. **65**, 255–260 (2012)
86. Millis, M.G.: First interstellar missions, considering energy and incessant obsolescence. J. Brit. Interplanet. Soc. **63**, 434–443 (2010)

87. Moore, J.H.: Kin-based crews for interstellar multi-generational space travel. In: Kondo, Y., Bruhweiler, F.C., Moore, J., Sheffield, C. (eds.) Interstellar Travel and Multi-Generation Space Ships, pp. 80–88. Apogee, Burlington (2003)

88. Mori, K.: Beamed propulsion by gravitational waves. J. Brit. Interplanet. Soc. **64**, 396–400 (2011)

89. Nakagaki, T., Yamada, H., Toth, A.: Intelligence: maze-solving by an amoeboid organism. Nature **407**, 470 (2000)

90. NASA: NASA's Kepler discovers its smallest 'habitable zone' planets to date. http://www.nasa.gov/mission_pages/kepler/news/kepler-62-kepler-69.html (2014). Accessed 7 Jan 2014

91. Nechaev, A.P., Polyakov, V.V., Morukov, B.V.: Martian manned mission: What cosmonauts think about this. Acta Astronaut. **60**, 351–353 (2007)

92. Obousy, R.: Project Icarus: A theoretical design study for an interstellar spacecraft. In: Johnson, L., McDevitt, J. (eds.) Going Interstellar, pp. 219–232. Baen, Riversdale (2012)

93. Obousy, R.K.: Project Icarus: a review of interstellar starship designs. J. Brit. Interplanet. Soc. **65**, 225–231 (2012)

94. Obousy, R.K., Tziolas, A.C., Long, K.F., Galea, P., Crowl, A., Crawford, L.A., Swinney, R., Hein, A., Osborne, R., Reiss, P.: Project Icarus: progress report on technical developments and design considerations. Brit. Interplanet. Soc. **64**, 358–371 (2011)

95. Obousy, R.K.: Project Icarus: A 21st century interstellar starship study. J. Brit. Interplanet. Soc. **65**, 325–329 (2012)

96. O'Rourke, D.H.: Genetic considerations in multi-generational space travel. In: Kondo, Y., Bruhweiler, F.C., Moore, J., Sheffield, C. (eds): Interstellar Travel and Multi-generation Space Ships, pp. 89–99. Apogee, Burlington (2003)

97. Peldszus, R., Dalke, H., Pretlove, S., Welch, C: The perfect boring situation-addressing the experience of monotony during crewed deep space missions through habitability design. Acta Astronaut. **94**, 262–276 (2014)

98. Petigura, E.A., Howard, A.W., Marcy, G.W.: Prevalence of Earth-size planets orbiting Sun-like stars. PNAS **110**(45), 1–6. (http://www.pnas.org/content/early/2013/10/31/1319909110)

99. Quarta, A.A., Mengali, G., Aliasi, G.: Optimal control laws for heliocentric transfers with a magnetic sail. Acta Astronaut. **89**, 216–225 (2013)

100. Regis, E., Jr.: The moral status of multigenerational interstellar exploration. In: Finney, B.R., Jones, E.M. (eds.) Interstellar Migration and the Human Experience, pp. 248–259. University of California Press, Berkeley (1985)

101. Rosen, E. (translator): Kepler's Somnium: The Dream, or Posthumous Work on Lunar Astronomy. Dover, Mineola (1967)

102. Sandal, G.M., Manzey, D.: Cross-cultural issues in space operations: a survey study among ground personnel of the European Space Agency. Acta Astronaut. **65**, 1520–1529 (2009)

103. Schwartz, D.W.: The colonizing experience: a cross-cultural perspective. In: Finney, B.R., Jones, E.M. (eds.) Interstellar Migration and the Human Experience, pp. 234–246. University of California Press, Berkeley (1985)
104. Schwartz, P.: Starships and the fates of humankind. In: Benford, J., Benford, G. (eds.) Starship Century: Toward the Grandest Horizon, pp. 27–39. Microwave Sciences/Lucky Bat, Charleston (2013)
105. Semyonov, O.G.: Radiation hazard of relativistic interstellar flight. Acta Astronaut. **64**, 644–653 (2009)
106. Semyonov, O.G.: Kinematics of maneuverable relativistic starship. Acta Astronaut. **75**, 85–94 (2012)
107. Sheffield, C.: Fly me to the stars: Interstellar travel in fact and fiction. In: Kondo, Y., Bruhweiler, F.C., Moore, J., Sheffield, C. (eds) Interstellar Travel and Multigeneration Space Ships, pp. 20–28. Apogee, Burlington (2003)
108. Shostak, S.: Searching for smart life around small stars. Astronomy **42**(2), 28–33 (2014)
109. Smith, E.E.: The Skylark of Space: Commemorative Edition. Bison/University of Nebraska Press, Lincoln (2001)
110. Stanic, M., Cassibry, J.T., Adams, R.B.: Project Icarus: analysis of plasma jet driven magneto-inertial fusion as potential primary propulsion driver for the Icarus probe. Acta Astronaut. **86**, 47–54 (2013)
111. Strong, J.: Flight to the Stars. Hart, New York (1965)
112. Stuster, J.: Behavioral Issues Associated with Long-Duration Space Expeditions: Review and Analysis of Astronaut Journals. Experiment 01-E104 (Journals): Final Report. NASA/TM-2010-216130. NASA/Johnson Space Center, Houston (2010)
113. Stratmann, H.: Chapter 7: Suspended animation: putting characters on ice. In: Medicine Meets Science Fiction. Springer, New York (in press)
114. Swinney, R.W., Long, K.F., Hein, A., Galea, P., Mann, A., Crowl, A., Obousy, R., Tziolas, A.C.: Project Icarus: exploring the interstellar roadmap using the Icarus Pathfinder and Starfinder probe concept. J. Brit. Interplanet. Soc. **65**, 244–254 (2012)
115. Thomason, S.G.: Language change and cultural continuity on multi-generational space ships. In: Kondo, Y., Bruhweiler, F.C., Moore, J., Sheffield, C. (eds.) Interstellar Travel and Multi-Generation Space Ships, pp. 100–103. Apogee, Burlington (2003)
116. Tomi, L., Kealey, D., Lange, M., Stefanowska, P., Doyle, V.: Cross-cultural training requirements for long-duration space missions: results of a survey of International Space Station astronauts and ground support personnel. Paper delivered at the Human Interactions in Space Symposium, 21 May 2007, Beijing (2007)
117. Underwood, D.R., Jones, B.W., Sleep, P.N.: The evolution of habitable zones during stellar lifetimes and its implications on the search for extraterrestrial life. Int. J. Astrobiol. **2**, 289–299 (2003). doi:10.1017/S1473550404001715

118. University of California Museum of Paleontology: Introduction to the "slime molds". http://www.ucmp.berkeley.edu/protista/slimemolds.html (2009). Accessed 30 Dec 2013
119. Wikipedia: Epsilon Eridani. http://en.wikipedia.org/wiki/Epsilon_Eridani (2013). Accessed 28 Dec 2013
120. Wikipedia: Habitability of orange dwarf systems. http://en.wikipedia.org/wiki/Hability_of_orange_dwarf_systems (2013). Accessed 7 Jan 2014
121. Wikipedia: Slime mold. http://en.wikipedia.org/wiki/Slime_mold (2013). Accessed 28 Dec 2013
122. Winterberg, F: Advanced deuterium fusion rocket propulsion for manned deep space missions. J. Brit. Interplanet. Soc. **62**, 386–401 (2009)
123. Winterberg, F.: Deuterium-tritium pulse propulsion with hydrogen as propellant and the entire space-craft as a gigavolt capacitor for ignition. Acta Astronaut. **89**, 126–129 (2013)
124. Woodcock, G.R.: To the stars! In: Schmidt, S., Zubrin, R. (eds.) Islands in the Sky: Bold New Ideas for Colonizing Space, pp. 183–197. Wiley, New York (1996)
125. Youngquist, R.C., Nurge, M.A., Starr, S.O., Koontz, S.L.: Thick galactic cosmic radiation shielding using atmospheric data. Acta Astronaut. **94**, 132–138 (2014)
126. Zubrin, R.: On the way to starflight: the economics of interstellar breakout. In: Benford, J., Benford, G. (eds.) Starship Century: Toward the Grandest Horizon, pp. 83–101. Microwave Sciences/Lucky Bat, Charleston (2013)

36556823R00086

Made in the USA
Lexington, KY
25 October 2014